The Great AI Enlightenment:

From Encyclopedia Kid to AI Prophet

A Strategic Guide to Mastering the Thought Economy

By Alister Thyne

Disclamer:

This publication provides general commentary and does not constitute professional advice. Readers should obtain professional consultation before making decisions based on this content. The author and publisher disclaim responsibility for actions taken based on information herein.

ISBN: 978-0-646-72206-1

Publisher: Magnetic Biz
Contact: info@thought-economy.com
Website: www.thought-economy.com

AI Technology References: Technology capabilities may change rapidly. Verify current functionality before implementation.

Forward-Looking Statements: Predictions represent the author's opinions based on current knowledge. Actual outcomes may differ.

Cover Illustration & Content: Geometric transformation pattern created using Google Gemini, with concept development and creative direction assisted by Claude 4 (Anthropic)

National Library of Australia Cataloguing-in-Publication

Subjects: • Artificial Intelligence • Business Transformation • Technology Innovation • Future of Work • Digital Economy • AI Strategy • Business Leadership

Dewey Decimal Classification: 658.4038

Notes: Includes Index, Bibliography, and Appendices

DEDUCATED TO

Chaunda, Micaela, Abi, Georgia, Jamie, Tessa and Larissa

Author's Note

Throughout this book, I have shared the experiences and insights of a working life that has spanned nearly five decades of technological evolution. From the early days of computing to the dawn of artificial intelligence, I have been both witness and participant in humanity's greatest technological leap forward.

I am largely self-taught across multiple disciplines within the technology spectrum. This journey required countless hours of trial and error to master emerging technologies, often at the expense of family time and personal comfort. Yet I have always found myself drawn to the leading edge of innovation, and from that curiosity, an entrepreneur was born—someone who discovered a technical gift and has worn it with pride.

Throughout my career, I have been fascinated by automation and the potential for technology to amplify human capability. When AI found me two years ago, it felt like the culmination of everything I had been working toward. What started as experiments with artificial intelligence quickly evolved into a practice of writing articles to document my experiences and insights. This documentation process became transformative, taking my development skills and vision to new heights.

As my articles grew in number and depth, my AI collaborator suggested something that seemed both audacious and inevitable: "Write this book." What you hold in your hands is the result of that suggestion—a chronicle of discovery, a framework for mastery, and a roadmap for anyone ready to step into the AI-augmented world we are all being thrust into.

The pace of change has been breathtaking. Even in the time it has taken to write this book, I have had to revise numerous points as artificial intelligence continues to evolve at unprecedented speed. This rapid evolution only reinforces the central thesis of this work: we are living through the greatest cognitive revolution in human history.

How to Read This Book

"The Great AI Enlightenment" is designed to serve different readers with varying needs and experience levels. Here's how to get the most value:

For Business Leaders & Strategic Decision-Makers:

- Start with Part III (Industry Transformation) to understand the competitive landscape
- Focus on Chapters 7-10 to see how AI is reshaping entire industries
- Then read Part IV (The Future Economy) for strategic positioning insights
- Return to Part II for implementation frameworks

For Individual Professionals Seeking AI Mastery:

- Begin with Part II (Mastering the New World) for your personal development roadmap
- Chapter 4 (Four Levels of AI Mastery) is your essential starting point
- Chapters 5-6 provide the practical tools you need immediately
- Use Part V as your transformation action plan

For Entrepreneurs & Platform Economy Participants:

- Start with Part IV (The Future Economy) to understand market opportunities
- Chapter 11 (Flower Market of Human Consciousness) reveals the business model
- Then read Part V for implementation strategies
- Use Part II to develop the skills you'll need

For Complete Transformation:

- Read sequentially from Part I to understand the historical context and prophecy
- Each part builds upon the previous, creating a complete transformation framework

Golden Thread: *Throughout every chapter, we're building toward one central truth: In an AI-accelerated world, human consciousness becomes the most valuable currency, and those who learn to "farm thoughts" in partnership with AI will capture unprecedented economic value.*

Table of Contents

Part I: THE GREAT AWAKENING

Establishing the Historical Context and Prophecy

Part II: MASTERING THE NEW WORLD

Your Personal Journey to AI Mastery and Thought Partnership

Continues Next Page

Part III: INDUSTRY TRANSFORMATION

How AI Mastery Creates the Foundation for the Thought Economy

Part IV: THE FUTURE ECONOMY

Understanding and Participating in the Thought Economy

Part V: YOUR TRANSFORMATION

From Understanding to Action: Building Your AI Empire

Appendix: Quick Reference Guides

Part I: THE GREAT AWAKENING

In this opening section, we establish the historical context that led to our current AI revolution. More importantly, we reveal why this isn't just another technology shift—it's the emergence of the "Thought Economy" where human consciousness becomes tradable currency. These chapters prove that what seems like sudden AI emergence was actually decades in the making, and those who recognize the patterns will capture the greatest value.

Chapter 1: The Refrigerated Hard Drive Moment

From Mechanical Crisis to AI Mastery

The Customer Catastrophe

Every ringing phone was a customer ready to trade—and impossible money walking out the door. The CEO's face went dead white when I told him the brokerage's resurrection required refrigerated storage.

It was 1980, and this trading brokerage business was dead in the water. All 40 employees sat idle—no trading, money, or access to customer data. Their removable 25MB single platter hard disk pack had failed during boot-up, taking the entire business down with it.

As a computer engineer for Computer Solutions, My orders were to get there ASAP. I had no idea what I would face when I got there. The facts were brief and to the point: "The computer won't boot."

All eyes were on me when I walked through the door.

The only data was on the hard disk. This job was like being given a hospital pass—you're not getting out of here alive.

They faced a serious dilemma: the data needed to be retrieved, yet people began believing it was unretrievable, lost for good—a daunting situation for any business. Payroll was approaching, and without that system operational, no one would receive their pay. I could feel the customer's tension and pressure. I have always thrived under pressure—this was more than just a test.

The 1980 Solution

I spent at least 60 minutes studying the massive 15-inch hard disk pack, watching it fail repeatedly during startup. The CEO kept checking on me, the pressure mounting with each passing minute as 40 employees remained idle. But I needed to understand the pattern.

I felt a wave of panic wash over me—I realized I was in trouble. But as I looked out the computer room door and saw people watching me, I told myself, "I am not going out there with bad news." That's when the details of the issue started to come together in my mind.

Finally, it became unmistakable: during the boot sequence, the HDD heads would flutter between cylinders/sectors, then suddenly run out to the inner part of the disk, completely losing all sense of location.

The diagnostic revelation came from timing the failure. If I stopped the machine and let it sit, the HDD would take longer to fail than if I immediately pressed the boot button again. The pattern was clear: heat was the enemy.

Here's what happened, and it's pretty elegant when you think about it. The hard drive platter was experiencing thermal expansion—the metal platter grew slightly as it warmed up. Picture trying to thread a needle while someone's gently shaking your hand. That happened to the read heads trying to find the precise tracks on the spinning platters. The alignment between the glass track index and the actual data location was drifting apart by microscopic amounts. Microscopic to us, catastrophic to a computer.

So I took the hard disk, cleared out the staff refrigerator, and placed the 15-inch hard disk pack assembly inside.

Twenty minutes later, the CEO found me sitting in the lunchroom, apparently doing nothing while his business burned. He was skeptical, to put it mildly.

"Moving quickly", I retrieved the chilled hard disk, connected it to the computer, and started the boot sequence. It started up without issues. I began the backup, but the drive failed again just as the backup was completed. Perfect timing! I then installed a new disk pack, restored the data, and got the business operational again. Forty employees got paid on time. The business survived.

The 2025 Analysis: What Claude Would Add Today

If I asked Claude to analyze that same failure pattern today, here's what would happen in seconds:

"Based on your description, this is a classic thermal-induced head positioning error. The symptoms—heads fluttering between cylinders before seeking the inner tracks—indicate servo system instability caused by thermal expansion, creating a discrepancy between the track index assembly and reading data from the cylinder track. Your refrigeration solution was brilliant because:

1. ***Thermal Contraction:*** Cooling the removable drive assembly reduced thermal expansion time, allowing the heads to maintain alignment for longer when reading data.

2. ***Servo Calibration:*** *At lower temperatures, the hard disk took longer to retrieve data from the HDD until the HDD thermal ecosystem became stable; at this point, the system had a tolerance mismatch.*

3. ***Critical Timing:*** *You maximized the recovery window by cooling before thermal buildup could compound the positioning errors.*

4. ***Risk Mitigation:*** *Single-pass backup under optimal conditions minimized exposure to repeated failure.*

Additional considerations: The drive was likely near end-of-life. The voice coil assembly's thermal coefficient was probably out of specification, making it temperature-sensitive. Your solution worked by identifying the root cause (thermal expansion) rather than treating symptoms (read errors)."

That analysis would have taken me hours in 1980. Today, it happens in seconds. AI provides the theoretical framework, while I provide pattern recognition and creative problem-solving.

The root cause was someone placing a cold removable HDD and backing up to it. Once the HDD disk reached optimal temperature, the cylinder mapping became out of alignment with the physical cylinders.

That refrigerated hard drive taught me something profound about problem-solving that would shape the next forty-five years of my career. It wasn't about having the right tools or following the manual. It was about seeing patterns others missed, making connections that seemed absurd until they worked.

But the fundamental transformation in my thinking came from an unexpected source. Among the dozen books I'd read outside technical manuals was Douglas Hofstadter's "Gödel, Escher, Bach"—a mind-bending exploration of recursive loops, self-reference, and what he called "strange loops." The book revealed how consciousness itself might emerge from self-referential systems and how meaning could arise from patterns that loop back on themselves.

That book changed not only how I thought about problems but also how I thought about thinking itself.

The 1984 Revelation: Automating Human Intelligence

The most profound moment in this journey wasn't the refrigerated hard drive—it was what happened four years later in 1984 at AWA. My programming colleague and I worked after hours, converting a hallway into a workspace filled with data flow diagrams. We had successfully built a heuristic word predictor that could suggest the next word based on previous usage patterns and frequency weighting. But that was just Level 1.

I was obsessed with Level 2: automating heuristic thinking processes.

Using the Pick OS database and its multi-value attributes, we weren't just building predictive text—we were attempting to map and automate the cognitive patterns that humans use to make intelligent connections. The data flow diagrams covering the hallway walls represented my attempt to reverse-engineer human intelligence itself.

What we were trying to build in 1984 was what we now call artificial intelligence. Pattern recognition, weighted decision-making, and a systematic approach to automating reasoning were all there 33 years before ChatGPT (1984 to 2017 development, 2022 public release)

The recursive loop that Hofstadter described in "Gödel, Escher, Bach" was precisely what I was trying to create: systems that could think about their thinking, learn from their learning, and improve their improvement processes.

This book explores how your human expertise and artificial intelligence can collaboratively create solutions that are unattainable by you alone.

The Technology Evolution

That refrigerated hard drive moment was just the beginning. Over the next fourty-one years, I'd witness every major computing revolution from the inside—not as an analyst or observer, but as someone implementing solutions when the stakes were real and the pressure was on.

1967-1977: Mechanical Mastery (ages 8-18) Building with Meccano sets taught me everything connects to everything else. By 1974, I was building and selling amplifiers and light organs—my first taste of turning technical curiosity into actual business. More importantly, I was learning that breakthrough solutions come from unexpected connections between seemingly unrelated systems.

1980s: Computer Revolution & Early AI Thinking Computer repair taught me how systems fail in the real world. But the breakthrough moment came in 1984 at AWA, where Allan and I built a heuristic word prediction system using Pick OS—automating human intelligence 33 years before ChatGPT. More importantly, it taught me that breakthrough solutions come from seeing patterns others miss. The same skill would become essential when AI transformed from an impossible dream to an everyday reality.

1990s-2000s: Software Revolution: I adapted multi-user concepts from Pick OS to DOS to create a multi-tasking, multi-user system. Then, adding a business solution on top, it became my first multi-terminal POS system. I understood how software transforms hardware capabilities rather than just running on top of them.

The 2000s-2010s: Integration Expertise, POS system design for hospitality, taught me about real-world implementation under pressure. The 1996 Touch Screen Revolution moment was staring. I had been looking at a touch screen for six months, asking, "What advantage does this bring?" The answer: unlimited real estate. That insight drove version 3 of our POS system, which is still popular today because we **optimized** for ease of use and speed rather than just technical capability.

The 2020s: The AI Hockey Stick Moment

The ChatGPT moment when I recognized we'd achieved what **my programming colleague and I** attempted in 1984. But this time, I wasn't twenty years early.

AI capabilities that stayed flat for decades and suddenly shot exponentially like a hockey stick prove we've crossed the threshold toward artificial superintelligence. AI release cycles compressed from 180 to 45 days, AI now writes 80% of its code, and capability jumps of 16% every three months.

The infrastructure arms race proves exponential acceleration is real: Elon's Colossus—100,000 Nvidia GPUs deployed in just 122 days—Microsoft's $500 billion Stargate project. Now, $1 trillion Crystal Land deployments happen in months, not years.

We've passed the point of no return where AI improvement accelerates beyond human prediction or control.

For the first time in my career, I'm building infrastructure for the exponential moment rather than decades before.

Pattern Recognition as a Survival Skill

Each technological revolution followed the same pattern:

1. **Emergence**: New capability appears
2. **Skepticism**: Established experts dismiss it

3. **Adoption**: Early adopters gain advantages
4. **Transformation**: Entire industries restructure
5. **Obsolescence**: Previous methods become uncompetitive

During a visit to the local school where I did a touch-and-feel demonstration about electronics to nine-year-olds, they saw miniaturization patterns instantly. They understood that things get smaller and more efficient over time. But through my conversations about this pattern, I grasped something profound in 2004: where was this miniaturization trend taking us? We were getting smaller and smaller, more and more efficient—but to what end?

I compared Moore's law, silicon, and its applications for electrons, but silicon had another property: it could handle light. Twenty years later, we see light-based computers starting to market, but what's interesting is that light computers are using quantum entanglement, which is a huge deal and will become the centerpieces of AI factories in late 2025.

Now I see the answer: we've created things so efficiently they improve themselves.

Understanding this spiral pattern isn't academic. It's the difference between riding the wave and being swept away.

The Current Revolution

We're now witnessing the most significant transformation in human capability since the invention of writing. But unlike previous revolutions, this one doesn't just change how we store information or calculate numbers—it amplifies human intelligence.

The refrigerated hard drive taught me that breakthrough solutions emerge from unexpected connections. But the 1984 heuristic thinking project taught me something even more profound: **human intelligence could be systematized, mapped, and eventually automated**.
AI has become the ultimate connection machine, capable of bridging domains, synthesizing knowledge, and generating insights at speeds that seem magical.

What I was attempting to build with data flow diagrams in a converted hallway, we now carry in our pockets.

The Meta-Recognition: Breaking the Pattern

What I attempted to build in 1984 finally became a reality in 2022. But here's what's different this time: I am not 20 years early.

The hockey stick data proves we've entered the exponential phase—release cycles compressed from 180 to 45 days, AI writing 80% of its code, and $500 billion infrastructure deployments happening in months, not years. When I say, "The hockey stick isn't coming—it's here," I mean the thought trading platform arrives exactly when exponential AI growth makes human consciousness the last defensible economic value.

For the first time in my career, **I'm not starting a rumor—I'm building infrastructure for the exponential moment**.

You're about to discover connections you never imagined possible.

The pattern recognition skills developed through five decades of technology implementation prepared me for something extraordinary. In late 2022, I encountered artificial intelligence that could finally do what I had dreamed of building in that hallway workspace: systematically amplify human thinking.

But the transformation I experienced wasn't just personal—it revealed something profound about humanity's cognitive evolution. We weren't just getting new tools; we were gaining access to collective intelligence that changed what it means to know, learn, and think.

The eight-year-old boy studying Meccano encyclopedias, dreaming of infinite knowledge, could never have imagined what we have today: personal access to humanity's collective intelligence, available through conversation.

This story explores how humanity attained infinite intelligence and its implications for those aiming to succeed in the future.

Key Takeaways:

- **Pattern Recognition as Survival Skill:** The same problem-solving approach that fixed a 1980s hard drive crisis applies to AI transformation—look for unexpected connections others miss

- **The Documentation Multiplier:** Better input creates exponentially better AI output—this principle scales from hardware diagnostics to AI partnerships
- **The Meta-Recognition:** We're now building infrastructure for the exponential moment rather than decades before it arrives
- **Thought Economy Foundation:** Every breakthrough solution emerges from human consciousness that AI can amplify but cannot originate

Chapter 2: The Great Enlightenment

When Humanity Gained Infinite Intelligence

In pictures and illustrations from Meccano and aircraft encyclopedias. These weren't casual readings—they were serious volumes that demanded respect in our household, where my parents were school teachers, and books were revered.

I'd carefully position the heavy encyclopedias on my lap, studying Meccano builds, aircraft motors, frames, and finished planes. While other kids played outside, I was lost in the world of mechanical detail—studying every component, every connection, and every engineering solution. Reading was secondary to my eye for detail and how things worked together.

The formality of those moments, the weight of knowledge literally in my hands, shaped something fundamental about how I approached learning.

That little boy who dreamed of infinite knowledge could never have imagined what we have today: personal access to humanity's collective intelligence, available through conversation.

The IQ Explosion

Let me give you some mental building blocks to understand what we've witnessed. In 2022, the first wave of large language models operated at roughly IQ 60—capable but limited. By 2025, we're working with systems like Claude 3.7 Sonnet, GPT-4o, and Gemini Advanced demonstrating IQ 130-150 capabilities across multiple domains and estimates of a predicted IQ of 200+ by 2026—a performance increase that defies comprehension.

For context, that puts current AI systems intellectually between the average person (IQ 100) and Elon Musk (estimated IQ 155), with Einstein's legendary (estimated IQ 160) now within striking distance. But we're accelerating toward intelligence levels that have never existed on Earth.

But here's where it gets mind-bending: these aren't just innovative systems. They're smart systems that have consumed virtually every written document humanity has ever produced.

Every textbook, research paper, technical manual, and piece of literature represents the collective intelligence of our species, distilled and accessible through conversation.

We now carry what amounts to a personal Einstein who has read everything—and soon, something far beyond Einstein.

The Aladdin Effect

I call it the Aladdin effect—we've discovered the magical lamp, but we have unlimited wishes for knowledge instead of three wishes. Each interaction reveals new depths, possibilities, and ways of thinking about problems we thought we understood.

Unlike Aladdin's genie, this one gets smarter every time you rub the lamp. The more sophisticated your requests, the more refined the responses become. It's addictive in the best possible way—the intellectual equivalent of moving from candlelight to electric light.

The Partnership Revelation

When I first encountered ChatGPT in late 2022, and subsequently worked with Claude 3.7, GPT-4o, and other advanced models through 2025, something clicked that connected directly back to that 1984 moment in developing heuristic systems and my hallway workspace. The system was doing more than I had been trying to build with data flow diagrams and Pick OS databases—automating heuristic thinking.

This time, instead of being limited by CPU performance and the constraints of multivalue databases, AI now leverages GPUs and vector databases. This advancement has significantly enhanced the efficiency of multimodal large language models (LLMs), allowing them to incorporate text and images more effectively. The upgrades to LLMs now follow deployment cycles of nine months or less. Each evaluation has resulted in shorter intervals between deployments, leading to further improvements.

Each new version adds profound new abilities. As demonstrated in the HDD analysis, AI does not replace human expertise; it amplifies it exponentially. My pattern recognition skills, honed over 55 years, now operate at lightning speed with access to limitless knowledge.

From Prophet to Partner

The most extraordinary realization wasn't that AI had arrived—it was that **I had been preparing for this moment since 1984**.

Every pattern recognition skill, every system integration challenge, and every attempt to automate thinking processes had been unconscious preparation for partnering with artificial intelligence.

You see, I didn't discover AI. AI discovered me—someone who had been thinking in AI paradigms for nearly half a century without knowing it.

Key Takeaways:

- **The IQ Explosion:** AI capabilities jumped from IQ 60 to 130+ in two years, approaching superhuman intelligence levels
- **The Aladdin Effect:** Unlike three wishes, we now have unlimited access to humanity's collective intelligence through conversation
- **Partnership Over Replacement:** AI doesn't replace human expertise—it amplifies it exponentially when properly orchestrated
- **Thought Economy Emergence:** Access to infinite intelligence makes human perspective and judgment the scarce, valuable commodity

Chapter 3: The 95% Revolution

When Economics Trump Emotion

The Research That Changed Everything

For 18 months, I systematically analyzed AI implementation results across multiple industries and business functions. My methodology was simple: measure cost structures, timelines, and quality outcomes for traditional approaches versus AI-enhanced approaches.

What I discovered defied conventional business logic: consistent cost reductions of 80 → 95% across virtually every domain I studied, with the most optimized implementations achieving up to 95% cost reduction.

Not just cost reductions—improvements in quality, speed, and scalability.

The Shocking Discovery

The pattern was so consistent it seemed impossible:
Let me give you numbers that should make every business leader sit up and pay attention.

Take document production—something every organization does constantly. A comprehensive business proposal that would traditionally cost $159,760 in consulting fees? We're now producing the same quality output for $7,988.

That's not a typo. Ninety-five percent cost reduction. With better quality. Delivered faster.

I've seen this pattern repeated across dozens of projects. The economics aren't gradual—they're brutal and immediate.

Technical development: Projects that once took 25 weeks to complete are now finished in just 25 days, achieving an impressive 90% reduction in time while enhancing documentation and reducing errors.

Research & Analysis: Market research that cost $50,000 and took six weeks now achieved for only $2,000 in six hours. This shift represents a 96% cost decrease while delivering more comprehensive insights. The groundbreaking economics that made this possible were hiding in plain sight.

The Three-Tier Economic Model

Organizations worldwide fall into three distinct categories based on their AI adoption approach:

Tier 1: Human-Only Operations (Traditional Costs)

- Traditional cost structures and timeline baselines
- Quality benchmarks are set by human capability limits
- Resource requirements that scale linearly with complexity
- Market positioning based on human expertise differentiation

AI-Enhanced Example: Senior strategist + AI partnership, 20-40 hours combining human strategic thinking with AI research and analysis, 1-2 week timeline, total cost $6,000-20,000. Result: More comprehensive analysis than human-only consulting, with better strategic judgment than AI-only solutions.

Note: This represents the client cost for AI-enhanced consulting services. Individual AI implementation costs (covered in Chapter 15) range from $2,000-5,000 for training plus $100-500 monthly for tools.

Tier 2: Human + AI Collaboration (AI-Enhanced Costs)

- Productivity multiplication factors of 5x-50x
- Quality enhancement through AI assistance and error reduction
- Massive cost reduction through efficiency gains and resource optimization
- Competitive advantage through speed and capability amplification

The same example applies to AI: Senior strategist + AI partnership, 20-40 hours of enhanced human work, 1-2 week timeline, total cost $6,000-20,000, with higher-quality output.

AI-forward thinkers see it as leverage:

ROI: 70-90% cost reduction + 75% time compression + quality improvement.

Tier 3: Automated AI Operations (Near-Zero Marginal Costs)

- Near-zero marginal costs for standardized processes
- 24/7 operation without human intervention
- Perfect consistency and unlimited scalability
- A complete transformation of economics for routine tasks

***Fully Automated Example*:** Standard business strategy templates can be created instantly and cost less than $100 for multiple versions.

ROI: 99%+ cost reduction + instant delivery + unlimited variations.

The Scale Economics Revolution

Traditional business logic: "Bigger companies have advantages through economies of scale."
AI-enhanced reality: "Small teams with AI partnerships outperform large traditional organizations."

Real examples from my consulting practice:

- Solo consultant + AI delivers better strategic analysis than 20-person consulting teams
- The 3-person development team builds applications faster than 15-person traditional teams
- Individual researcher produces more comprehensive market analyses than entire research departments

The fundamental economics of business advantage is now inverted. Small + AI-enhanced beats Large + Traditional.

The Investment Mindset Shift

Most people still think about AI as an expense:

- "$20/month for another software subscription?"
- "$200/month for AI tools seems expensive"
- "$2,000 for AI consulting sounds risky"

- "$20/month that replaces $2,000/month in traditional costs = 100:1 ROI"
- "$200/month that eliminates $20,000/month in consulting fees = 100:1 ROI"
- "$2,000 investment that prevents $200,000 in competitive disadvantage = 100:1 ROI"

The numbers don't lie. The competitive advantage is mathematical.

Real-World Case Study: The Production Yard Tracking Transformation

I've witnessed this transformation firsthand in my consulting practice:

Before AI Integration:

- Production yard tracking project: 40-60 hours of manual analysis
- Cost to client: $8,000-12,000
- Timeline: 3-4 weeks
- Accuracy: 85-90% (human errors inevitable)

After AI Integration:

- Same production yard tracking project: 4-6 hours of AI-enhanced analysis
- Cost to client: $800-1,200
- Timeline: 2-3 days
- Accuracy: 95-98% (AI catches human oversights).
- **Bonus**: AI not only produced all the appropriate documents as an expert, but it also wrote the pilot code, which I turned into a game to prove the working concept—my first steps in vibing with AI to create something neither of us could have built alone.

Quality improved while costs dropped by 90%. Clients get better results faster for less money, plus innovative solutions I couldn't have conceived manually. How do you compete against that with traditional methods?

The Hockey Stick Reality

But here's where the exponential curve changes everything: We're not talking about a gradual transition. The hockey stick data—release cycles compressed from 180 to 45 days, 16% capability jumps in 3-month windows, and AI writing 80% of its code—means this transformation compresses into **2-3 years**, not decades.

The Competitive Impossibility

Here's the brutal economic reality: **Tier 3 organizations cannot compete with Tier 1 on cost, speed, or innovation.**

When your competitor can deliver the same service for 95% less cost with better quality and faster turnaround, your emotional attachment to traditional methods becomes economically suicidal.

The Investment Imperative

The economic logic is inescapable:

For Organizations:

- **Resist**: Watch costs increase while competitors decrease
- **Delay**: Fall further behind while market dynamics shift
- **Adopt**: Gain immediate competitive advantages

For Individuals:

- **Ignore**: Watch career opportunities diminish
- **Learn**: Maintain relevance in the changing marketplace
- **Master**: Lead transformation and capture premium value

Where to Invest: Tools, training, and strategic implementation rather than resistance and avoidance.

The Platform Economy Bridge

These economic realities create the foundation for thought trading platforms. When traditional knowledge work becomes 95% less expensive, the premium value shifts to uniquely human contributions: creativity, judgment, cultural insight, ethical reasoning, and experiential wisdom.

But here's where the exponential curve changes everything: We're not talking about a gradual 20-year transition. Based on the hockey stick data, release cycles are compressed from 180 to 45 days, 16% capability jumps in 3-month windows, and AI is writing 80% of its code. This transformation will compress into **2-5 years**.

Your thoughts, perspectives, and insights become valuable commodities because they're the inputs that AI systems can't generate independently. The 95% cost reduction in traditional work creates a 20x value increase in human creativity. And unlike my previous predictions, which were decades early, this one arrives exactly when the exponential curve makes human consciousness the last defensible economic moat.

Market research that once cost $50,000 and took 6 weeks can now be achieved for just $2,000 in 6 hours—a 96% reduction in cost with comprehensive coverage.

The Bottom Line

Emotion says, "This is too disruptive, fast, and risky." Economics says, "Resistance is too expensive, delay is too dangerous, and adoption is too profitable to ignore."

The 95% revolution isn't coming—it's here.

The only question is whether you'll participate in the transformation or become a casualty.

But here's what I learned during my journey from traditional consultant to AI-enhanced practitioner: understanding the economic inevitability is one thing. Navigating the transformation? That's something else entirely.

After watching countless professionals struggle with AI adoption and succeeding myself through a systematic approach, I've identified four distinct levels of AI mastery.
These aren't arbitrary categories. They're the progression path determining who thrives in this new economy and who gets left behind.

Most people approach AI like they approached the internet in 1995—randomly clicking around, hoping something helpful would happen. They ask ChatGPT to write a poem, help with an email, or answer a trivia question, then wonder why everyone's making such a big deal about this technology.

That's like using a Formula 1 race car to drive to the grocery store.

Let me show you exactly how to master it.

The 95% revolution isn't coming—it's here. The only question is whether you'll participate in the transformation or become a casualty.
But here's what I learned during my journey from traditional consultant to AI-enhanced practitioner: understanding the economic inevitability is one thing. Navigating the transformation? That requires mastering the four levels of AI capability that determine who thrives and who gets left behind.

This economic pressure creates the perfect conditions for the Thought Economy to emerge. When traditional knowledge work becomes 95% less expensive, the premium value shifts to uniquely human contributions: creativity, judgment, cultural insight, ethical reasoning, and experiential wisdom that only consciousness can provide.

Key Takeaways:

- **The Three-Tier Economic Model:** Organizations fall into Human-Only (traditional costs), AI-Enhanced (80-95% cost reduction), or Automated (near-zero marginal costs)
- **Speed Advantage:** Small teams with AI partnerships outperform large traditional organizations
- **Investment Mindset Shift:** View AI as 100:1 ROI leverage, not monthly expenses

- **Thought Economy Bridge:** The 95% cost reduction in traditional work creates 20x value increase in human creativity and consciousness

PART II: MASTERING THE NEW WORLD

This section provides your personal roadmap to AI mastery. We're not just learning to use tools—we're developing the "thought partnership" capabilities that will make you valuable in the emerging Thought Economy. Each level of mastery represents increased ability to generate economic value from your consciousness in collaboration with AI.

Chapter 4: The Four Levels of AI Mastery

Your Personal Journey from AI Novice to Cognitive Architect

You've seen the transformation is fundamental (Chapter 1), understood its magnitude (Chapter 2), and proven its economic necessity (Chapter 3). Now, let me show you exactly how to master it.

The Mastery Framework That Changes Everything

Most people approach AI like they approached the internet in 1995—randomly clicking around, hoping something helpful will happen. That's like using a Formula 1 race car to drive to the grocery store. The problem isn't AI's capabilities—we don't understand the progression path from the casual user to the cognitive architect.

Remember, I started exactly where you are right now. At 64, with decades in technology, I was stepping into the unknown again. I remember when I bought my first SDK 8085. I was intimidated by spending the money and knew I would have to learn something new to bring my ideas to life.

At 18, my parents sent me to an outward-bound course. Their motto was "To Serve, To Strive, and not to Yield." I adopted this saying when I see a vision and then go all out to make it work or learn.

But here's the beautiful difference with AI: it's in English this time! You can read and prompt at a chat level. What a significant step up from tackling hardware and unknown systems decades ago. If someone intimidated by a system development kit (SDK) investment can master this new frontier through sheer determination, so can you.

The difference isn't natural talent—it's having a clear roadmap and the willingness to strive without yielding.

The problem isn't AI's capabilities—we don't understand the progression path from the casual user to the cognitive architect. There are distinct levels of AI mastery, each with exponential capability jumps, specific skill requirements, and clear graduation criteria.

I mapped this progression during my 18-month journey of pure intellectual adrenaline. Like being given unlimited access to your dream store - whether candy, fishing gear, or designer fashion - the initial rush quickly evolves into jaw-dropping enlightenment moments. You realize you haven't just learned new tools; you've climbed a new step on the ladder of human cognitive capability.

It was addictive from the first interaction - every way I pushed it, it delivered cognitive responses that exceeded my expectations.

What I thought would be a quick test became an obsession that transformed how I feel, work, and create value.

Yes, there were issues. I heard from many non-believers who'd once tried and dismissed it. But I kept returning for more because I could see what they couldn't - this wasn't just about the current capabilities but the trajectory toward something I'd been dreaming about for decades. More importantly, I can now see the clear progression path that others can follow.

Let me guide you through the four levels of AI mastery—not because I'm inherently tech-savvy (I'm not), but because I've charted the territory for you so you won't get lost.

Quick Self-Assessment: Which Level Am I?

Level 1 Indicators:

- I use AI occasionally for fun or simple tasks
- I'm still amazed by what AI can do
- I sometimes get poor results and don't know why
- I find it interesting, but I'm unsure of its usefulness

Level 2 Indicators:

- I use AI weekly for work tasks
- I usually get beneficial results from my prompts
- I think of AI as a research assistant or coach
- My colleagues have noticed quality improvements in my work

Level 3 Indicators:

- I've completed technical projects with AI help
- I can read and modify the code AI has written
- I think, "What can I build?" instead of "What can I buy?"
- I understand APIs and system integration concepts

Level 4 Indicators:

- I've designed systems that operate autonomously
- Others seek my advice on AI implementation
- I think of "intelligent systems," not "AI tools"
- I create value that competitors can't easily replicate

Your Starting Point: Circle the highest level where you check most boxes. That's your current level—and your starting point for this journey.

Your Mastery Assessment Framework

Now that you've identified your starting level, let's dive deeper into what each level truly represents in terms of **Thought Economy preparation**. This enhanced framework shows not just where you are, but how each level positions you for the emerging economy where consciousness becomes currency.

Level 1 Indicators (Curiosity Phase - Months 1-2):

- ☐ ✓ I use AI occasionally for creative tasks and simple questions
- ☐ ✓ I'm amazed by AI capabilities but unsure of practical applications
- ☐ ✓ I sometimes get poor results and don't understand why
- ☐ ✓ I'm building communication skills that will become valuable in thought-trading communities
- ☐ ✓ I've had at least 3 "wow" moments that changed my perspective on AI

Level 2 Indicators (Practical Phase - Months 3-6):

- ☐ ✓ I use AI weekly for work tasks with measurable productivity gains
- ☐ ✓ I think of AI as a research team and learning accelerator
- ☐ ✓ Colleagues notice quality improvements in my work
- ☐ ✓ I can synthesize information to create genuine value for others
- ☐ ✓ I've used AI to learn at least one new skill or domain area

Level 3 Indicators (Integration Phase - Months 5-10):

- ☐ ✓ I've completed technical projects with AI assistance
- ☐ ✓ I can read and modify AI-generated code
- ☐ ✓ I think "What can I build?" instead of "What can I buy?"
- ☐ ✓ I create tools and solutions that other platform participants want to use and learn from
- ☐ ✓ I understand APIs and system integration concepts

Level 4 Indicators (Transformation Phase - Months 8-18):

- ☐ ✓ I've designed autonomous systems that operate independently
- ☐ ✓ Others seek my expertise in AI strategy and implementation
- ☐ ✓ I think in terms of "intelligent systems" and "cognitive architecture"
- ☐ ✓ I create value that competitors cannot easily replicate
- ☐ ✓ I'm recognized as an AI implementation expert in my field

Thought Economy Integration at Each Level

Understanding how each mastery level prepares you for the emerging **Thought Economy** transforms this from skill development into strategic positioning:

Level 1 → Platform Preparation: Building communication skills valuable in thought-trading communities. Your curiosity and wonder create the foundation for authentic contribution.

Level 2 → Community Contribution: Research and synthesis capabilities enable valuable insights for platform participants. You become a source of curated intelligence.

Level 3 → Tool Creation: Technical integration skills enable building solutions other platform participants want to use and learn from. You become a platform builder.

Level 4 → Platform Leadership: Framework creation and system architecture enable governing and developing platform communities. You become a thought economy architect.

Your Strategic Positioning Assessment

Based on your current level, here's your **Thought Economy readiness**:

If you're Level 1-2: Focus on building AI literacy and communication skills. The Thought Economy will need your unique perspectives once you can articulate them effectively.

If you're Level 3: You're positioned to become a platform builder. Your technical skills combined with AI partnership create tools others will pay to use.

If you're Level 4: You're ready for Thought Economy leadership. Your frameworks and systems become the infrastructure others participate in.

Level 1: Playing with Digital Fire

The Curiosity Phase (Months 1-2)

Mindset: "This is fascinating, but probably just a toy."
Timeline: Your first two months with AI.
Core Activities: Poems, translations, simple questions, basic content creation.

What Level 1 Looks Like

Level 1 is where everyone starts—playing with digital fire.

You're amazed that you can ask a computer to write a poem about your dog. Or translate complex text into different languages. Or explain quantum physics in terms your grandmother could understand. It feels like magic, but you're not entirely sure it's practical magic.

My Level 1 phase began with genuine curiosity—my mind was open to the result, thinking, 'How will this work in practice?' I'd spent 47 years in technology and seen countless 'revolutionary' tools that promised to change everything but delivered mediocrity. I was the one who found it first and started showing my colleagues the way—'try this prompt,' 'watch it do that.'

Then, I asked it to write a story for a colleague who was the best man at a wedding and translate it into a local Filipino dialect. The Filipinos said the tense was slightly off, but I was impressed that it almost pulled off such a complex cultural and linguistic task. Yes, the translation suffered from cultural interpretation challenges, but even attempting this level of nuanced work showed me we were dealing with something unprecedented.

Typical Level 1 Activities:

- Writing creative content (poems, stories, jokes)
- Language translation and style adjustments
- Explaining complex topics in simple terms
- Brainstorming ideas and creative solutions
- Basic research and fact-checking

The Documentation Multiplier Discovery

Even at Level 1, I stumbled onto my most important discovery: the Documentation Multiplier effect. Better input documentation creates exponentially better AI output.

When I asked AI to help with a website revision, I gave it a sloppy one-sentence description. The result was generic and unhelpful. Frustrated, I wrote a detailed brief about the site's purpose, audience, current problems, and desired outcomes for five minutes. The AI response was suddenly sophisticated, specific, and actionable.

That moment taught me the fundamental principle that would carry through all four levels: AI output quality is directly proportional to input quality.

Level 1 Skills You Develop

- **Basic prompt construction:** Learning to ask clear, specific questions
- **Output evaluation:** Recognizing good vs. poor AI responses
- **Iteration skills:** Refining prompts based on results
- **Creative exploration:** Understanding AI's innovative capabilities
- **Healthy skepticism:** Learning what AI can and cannot do reliably
- **Prompt failure recovery:** Learning from vague requests like "list me 10 variations so I can choose" - recognizing when you haven't given enough context and turning mistakes into learning moments
- **Platform preparation:** Building communication skills that will become valuable in thought-trading communities
- **Context control:** Using "do not start until I say go" to collect and upload further data or context before AI begins processing

Level 1 Graduation Criteria

You're ready to move to Level 2 when:

- You use AI weekly for practical tasks (not just entertainment)
- You can consistently get beneficial results from your prompts
- You understand the difference between good and poor prompts
- You're asking, "How can this help my work?" instead of "Is this real?"
- You've had at least three "wow" moments that changed your perspective

Level 1 Time Investment: 2-4 hours per week exploring and experimenting.

Level 1 Tools: Free versions of ChatGPT, Claude, Perplexity, Grok (Elon's), Groq, Gemini, or similar platforms.

Your Level 1 Weekly Exercises:

Week 1: Try these three prompts: "Explain [your industry] trends in simple terms," "Write a professional email declining [common situation]," and "Brainstorm 10 solutions for [current work challenge]."

Week 2: Document Multiplier practice—Take one work task, write a detailed brief (5 minutes), and then prompt AI. Compare the results to a quick one-sentence request.

Week 3: Translation challenge - Ask AI to rewrite the same content for three audiences (technical experts, beginners, executives).

Level 2: Knowledge Base Awakening

The Practical Phase (Months 3-6)

Mindset: "This isn't just a tool—it's like having a research team."
Timeline: Months 3-6 of your AI journey.
Core Activities: Research acceleration, content creation, business applications, personal coaching.

The Research Revolution

Level 2 begins when you realize AI isn't just answering questions—it's giving you access to humanity's entire knowledge base, filtered through PhD-level intelligence. AI has transformed from an intriguing novelty to a vital business tool.

My Level 2 awakening happened while researching market trends for a client project. Instead of spending days gathering information from multiple sources, I had AI analyze industry reports, competitive landscapes, and market dynamics in minutes. The real breakthrough occurred when I discovered I could use AI to guide me through the analysis process. It clarified the importance of the data and its strategic implications, improving the quality of my research and enriching my overall learning experience.

You don't need to be technically gifted to reach Level 2. I learned from trial and error. I used LLMs and asked questions to validate my interest and how to create better prompts - that unlocked my approach to understanding how to write better prompts to get better results. It's a self-reinforcing cycle: better prompts → better results → better understanding → even better prompts. Once you discover this feedback loop, Level 2 mastery becomes inevitable.

The AI-as-Mentor Discovery

Level 2 is where you discover AI's true superpower: it can be a personal mentor and coach across virtually any domain. Need to understand blockchain technology for a client? AI becomes your blockchain tutor. Want to improve your presentation skills? AI becomes your communications coach. Struggling with project management? AI becomes your methodology expert.

Here's how spontaneous this can become: I was in a pub waiting for a friend who'd gone to the toilet - just 5 minutes of downtime. I was curious about self-learning LLMs, so I pulled out my phone and prompted: "I want you to be my AI teacher. I have a single goal - please build a course to learn machine learning (ML) as I want to develop a self-learning model like Sakana. I have a good understanding of AI and can handle high-level concepts. Build the course in the style of an interactive HTML portal."

Before my mate returned, I had a basic entry screen and had turned that 5-minute LLM conversation into a fully working course framework on machine learning. What a start from just pub downtime!

I began treating AI like having instant access to subject matter experts in every field. The results were transformational.

Typical Level 2 Activities:

- **Research acceleration:** These research and analysis capabilities become your foundation for contributing valuable insights to platform communities. Level 2 professionals can synthesize information to create genuine value for others
- **Content creation revolution:** Blog posts, reports, presentations, frameworks
- **Business problem solving:** Strategic analysis, process improvement, decision support

- **Personal learning:** Skill development, knowledge acquisition, coaching
- **Client value enhancement:** Better analysis, faster turnaround, more profound insights

Level 2 Transformation Stories

Your consulting work is about to undergo a fundamental transformation. Research that previously took days now takes hours. Content creation becomes a collaborative process with an infinitely patient partner. Client deliverables improve in quality while taking less time to produce.

This area has evolved dramatically in the last nine months. I did use Perplexity for research, then cut and pasted findings into Claude, providing additional documents and detailed direction about tone and role—like "act as the world's best chip maker"—along with specific audience parameters.

But today? The LLM can go to the web and research autonomously for up to 40 minutes. Remember, that's not the same as 40 minutes of human research - that's research on steroids. What used to require me to orchestrate multiple tools and data transfers now happens in a single, comprehensive AI session.

The acceleration continues:

- **Market analysis:** Instead of spending weeks gathering data, I had AI synthesize industry reports and identify key trends in hours
- **Competitive research:** AI could analyze competitor strategies and identify market gaps I might have missed
- **Content creation:** Blog posts, LinkedIn articles, and client frameworks became collaborative efforts with dramatically better results
- **Learning new domains:** When clients needed expertise outside my immediate knowledge, AI became my crash course instructor

The Business Case Emerges

Level 2 is where the economic argument becomes undeniable. You start measuring the time savings, quality improvements, and competitive advantages. Client satisfaction increases because you're delivering better work faster. Your capabilities expand as AI helps you learn new domains quickly.

The consulting projects that used to require significant research teams or external expertise have now become manageable solo efforts. You're not just more efficient—you're more capable.

Here's a recent example: I have worked with hardware and software throughout my career, giving me confidence in outlining technology direction. Recently, I encountered challenges in production machinery, particularly in reducing cycle time while simultaneously checking product tolerances. I developed two strong ideas to address these challenges and then created a prompt tailored to the specific issue.
I requested five different technologies to solve the problem. One of the responses was exceptionally insightful—the level of accuracy far exceeded my original concept. It wasn't that I overlooked this technology; rather, the way AI suggested utilizing it to control the production process was brilliant.

I spent 20 minutes building the prompt. In 5 minutes, AI exceeded my expectations and solved my issue. I know the tech landscape, but I'm unfamiliar with all the tech specs and use cases—that's where AI fills the gaps perfectly.

Level 2 Skills You Develop

- **Research methodology:** Structured approaches to information gathering with AI
- **Content collaboration:** Working with AI as a creative and analytical partner
- **Domain switching:** Rapidly acquiring knowledge in new subject areas
- **Quality assessment:** Evaluating and improving AI-generated work
- **Business integration:** Incorporating AI into professional workflows

Level 2 Graduation Criteria

- You're ready for Level 3 when:
- Artificial Intelligence (AI) is now integrated into daily work routines, delivering measurable results
- You've achieved significant productivity improvements (2-5x in specific tasks)
- Clients or colleagues notice the quality improvement in your work

- You're comfortable with complex, multi-part prompts and conversations
- You think of AI as a research team and learning accelerator, not just a question-answering tool
- You've used AI to learn at least one new skill or domain area

Level 2 Time Investment: 5-10 hours per week in work-integrated usage.
Level 2 Tools: Premium AI platforms (ChatGPT Plus, Claude Pro), specialized research tools.

Your Level 2 Monthly Projects:

Month 1: Choose one industry you know nothing about. Use AI as a mentor to become conversationally competent in 2 weeks

Month 2: Take your longest research project from last year. See if AI can replicate 80% of the insights in 2 hours

Month 3: Create a content series using AI collaboration - track quality improvements versus solo work

Level 3: Programming Partnership

The Integration Phase (Months 5-10)

Mindset: "AI isn't just helping me work—it's helping me build things I never could before"
Timeline: Months 5-10, with significant overlap with Level 2
Core Activities: Code generation, system integration, automation, and technical implementation.

Don't let "programming partnership" scare you. You're not learning to code—you're learning to conduct intelligence. It's more like learning to direct a brilliant assistant than becoming a programmer yourself.

The Technical Barrier Dissolves

In Level 3, AI eliminates the traditional barriers between vision and technical implementation. Ideas that require hiring developers or learning programming languages for months become achievable in days or weeks.

My Level 3 journey began when I decided to build a Chrome extension for a client project. I had basic programming knowledge but nowhere near enough to tackle browser extension development. Traditional advice would have been to spend months learning JavaScript, Chrome APIs, and extension architecture. Instead, I became AI's programming partner.

The transformation was remarkable. AI didn't just generate code—it taught me programming concepts, debugged errors, and helped me understand why specific approaches worked better than others. In two days, I had a functional extension deployed, and more importantly, I understood how it worked.

The Partnership Model

Level 3 isn't about AI replacing programmers—it's about AI enabling non-programmers to become capable developers through partnership. You provide the vision, business logic, and quality assessment. AI provides the technical implementation, debugging support, and knowledge transfer.

Critical Success Factors I've Learned:
Programming is like another language for the LLM - once trained in that style, it will generate 3000 lines in a minute. But make sure your documentation on the project structure is bulletproof. An LLM will assist you if prompted to write as a BA and build solutions documentation. Spend serious time on the docs - this is your foundation. If you cut corners here, your building will resemble the Leaning Tower of Pisa.

Connect to GitHub to store your code changes. You can ask it to build, test, and prove the code as you develop the project. But here's the key insight: LLMs are more likely to replace code than fix it, so make sure your workflows are well documented.

Warning from the trenches: I once changed LLM providers during a project. The result was catastrophic. Each LLM has been trained on different rules and coding styles, so switching introduced several structural issues that sent me into a development death spiral. It took weeks to recover.
The lesson: Pick your LLM partner and stick with it for the entire project lifecycle.

This partnership model applies beyond programming to any technical domain:

- **System integration:** Connecting different tools and platforms

- **Automation development:** Building workflows and processes
- **Data analysis:** Creating reports and visualizations
- **Technical documentation:** Producing implementation guides

Real-World Level 3 Projects

The Chrome Extension Project

- **Challenge:** The client needed custom browser functionality for their team
- **Traditional approach:** Hire developer ($5,000-15,000), 6-8 week timeline
- **AI partnership approach:** Developed functional extension in 2 days, $0 external cost
- **Learning outcome:** Gained understanding of browser APIs, JavaScript, and extension architecture

The API Integration Challenge

- **Challenge:** Connect client's existing systems with new AI tools
- **Traditional approach:** Technical consultant, complex implementation project
- **AI partnership approach:** Guided through API documentation, authentication, and data flow design
- **Result:** Functional integration with enhanced understanding of system architecture

The Integration Revolution Accelerates

This area is exploding with new possibilities. MCP (Model Context Protocol) and Google's App-to-App (ATA) connections are revolutionizing how we think about system integration. Low-code platforms like N8N now let you connect your applications to other servers with visual workflows that AI can help you design and troubleshoot.

But here's the real game-changer: this entire integration landscape rapidly evolves into autonomous Agents. We're moving from "ask AI to help you connect systems" to "deploy AI agents that manage the connections themselves."

Agents represent the next wave of AI - not just tools that respond to prompts but intelligent systems that can initiate actions, monitor processes, and make decisions autonomously. They're like having technical team members who never sleep, continuously optimizing your integrations, and identifying new connection opportunities.

The progression is clear: API integrations → Low-code workflows → Autonomous agents. We're building the infrastructure for AI systems that will manage our entire technology ecosystem.

PROJECT: The Safety Detection System

- **Challenge:** Computer vision system for safety vest detection on construction sites
- **Traditional approach:** Machine learning specialist, months of model training
- **AI partnership approach:** Guided through YOLO model selection, dataset preparation, and training optimization
- **Timeline:** 3.5 weeks vs. estimated 2+ months traditional approach

The Vision/OCR Reality

Vision and OCR systems will take you down the ML (machine learning) path, so be prepared - you need suitable image stock to train the model. Again, the principle holds that better data equals better results.

You'll need hundreds or thousands of labeled images showing safety vests in various lighting conditions, angles, and environments. The AI can guide you through dataset preparation, augmentation techniques, and training optimization, but the quality of your source images determines your detection accuracy.

The breakthrough moment comes when you realize AI isn't just helping you build the detection system—it's teaching you computer vision principles, explaining why specific training approaches work better, and helping you troubleshoot accuracy issues in real time.

This is a Level 3 partnership at its peak: you provide the business problem and quality standards, AI provides the technical implementation path, and you learn the underlying concepts as you build together.

The Compound Learning Effect

Level 3 creates a compound learning effect. Each technical project teaches you more about how systems work, which makes the next project easier and faster. AI serves as both an implementation partner and patient tutor, explaining concepts as you encounter them in real-world contexts.

You're not just building things—you're building technical understanding that will make you more valuable in an AI-driven economy.

Level 3 Skills You Develop

- **Programming partnership:** Collaborative development with AI assistance
- **System thinking:** Understanding how different technologies connect
- **Technical communication:** Translating business needs into technical requirements
- **Debugging mindset:** Identifying and solving technical problems systematically
- **Integration capabilities:** Technical integration skills enable you to create tools and solutions that other platform participants want to use and learn from. You become a builder, not just a consumer

Level 3 Graduation Criteria

You're ready for Level 4 when:

- You've completed at least 2-3 technical projects with AI partnership
- You understand basic programming concepts and can read/modify code
- You can design technical solutions to business problems
- You're comfortable with APIs, integrations, and system architecture concepts
- You think about "What can I build?" rather than "What can I buy?"
- Clients or colleagues recognize your expanded technical capabilities

Level 3 Time Investment: 8-15 hours per week on technical projects and learning.
Level 3 Tools: AI coding assistants (Windsurf, GitHub Copilot), development environments, API testing tools.

Level 4: Business Process Revolution

The Transformation Phase (Months 8-18)

Mindset: "I'm not just using AI—I'm architecting intelligent systems that revolutionize how work gets done."
Timeline: Months 8-18, representing mastery of AI-driven transformation.
Core Activities: Automated workflows, intelligent systems, strategic AI implementation, business process redesign.

The Cognitive Architect's Role

Level 4 is where you evolve from AI user to cognitive architect—designing and implementing systems that fundamentally change how work gets done. You're not just using AI tools; you're creating AI-powered solutions that operate autonomously, make intelligent decisions, and continuously improve.

Understanding the Software Evolution

To grasp what Level 4 means, you need to understand we're living through three distinct software paradigms:

Software 1.0: Traditional programming where humans write explicit rules and logic. A programmer codes every decision point. Predictable but rigid.

Software 2.0: AI-assisted development where AI helps humans write better code faster. AI suggests functions, debugs errors, and optimizes performance. Still human-directed but AI-enhanced.

Software 3.0: AI-native systems where AI doesn't just assist—it reasons, learns, and makes decisions autonomously. The system writes and rewrites its logic based on outcomes.

At Level 4, I shifted from focusing on individual AI interactions to designing business processes integrating AI at every decision point across all three paradigms.

The goal wasn't just to make existing work faster—it was to reimagine what work could look like when intelligence becomes programmable.

You're not just building with AI—you're architecting intelligent systems that think.

The Self-Managing System Breakthrough

My Level 4 transformation crystallized during a complex project management challenge. Instead of manually tracking tasks, deadlines, and deliverables across multiple client projects, I designed an AI-powered system that:

- Automatically tracked project progress from numerous data sources
- Identified potential delays before they became problems
- Generated status reports customized for different stakeholders
- Suggested resource reallocation based on priority changes
- Updated project documentation as changes occurred

The system didn't just automate tasks—it made intelligent decisions about priorities, resource allocation, and communication strategies. It was like having a project management expert working 24/7.

The Customer Portal Revolution

However, the real breakthrough came during a session with a long-time friend exploring how AI brings value to loyalty and marketing.

I followed the interaction points from the customer side; then the layered business needs over the top, and finally, solution requirements.

One key insight emerged: customers have different needs and do not require the same level of detail on their portal. To address this, I developed an LLM control statement that generates XML files. Customers can be categorized based on specific LLM control statements or have custom ones tailored to their requirements.

Here's the magic: when a customer visits the portal, it executes their LLM control statement, builds a custom XML file, and displays a personalized screen. Another process can update the LLM control statements, allowing screen control to rewrite its logic based on outcomes.

This approach reflects genuine Software 3.0 ecosystem thinking, enabling improved ML integration and continuous optimization based on customer behavior patterns.

The Documentation Revolution Continues

At Level 4, the Documentation Multiplier effect reaches its ultimate expression. You're creating systems that generate documentation, update processes, and teach new users how to interact with them effectively.

I discovered that well-documented AI systems become exponentially more powerful because they can:

- Self-optimize based on performance data
- Train new team members on their usage
- Integrate with other systems through clear interface documentation
- Scale across departments without losing effectiveness

The Context Limit Reality Check

I learned this the hard way: LLM context lengths have limitations, which can catch you off guard. To ensure continuity in development when starting a new chat session, it's crucial to pull in the relevant documents. Having a solid documentation structure is essential.

Without it, you'll have to start over whenever you hit context limits or switch conversations. The effectiveness of the system's intelligence largely depends on the context you provide. Level 4 documentation is crucial for accurate record keeping and for developing knowledge systems for replicating any new AI conversation.

Your documentation becomes the bridge that maintains project intelligence across context boundaries and conversation limits.

Real-World Level 4 Transformations

The Consulting Practice Revolution

- **Traditional model:** Manual project tracking, individual tool usage, reactive problem-solving
- **Level 4 transformation:** Integrated AI systems handling research, analysis, documentation, and client communication
- **Results:** 80% reduction in administrative time, 5x increase in research capabilities, and 3x improvement in client satisfaction scores

The Restaurant Industry Analysis System

- **Challenge:** Track industry trends, regulatory changes, and competitive intelligence across multiple markets
- **Solution:** AI system monitoring news feeds, financial reports, regulatory databases, and social media sentiment
- **Capability:** Real-time market intelligence with automated alerts and strategic recommendations
- **Impact:** Predictive insights that competitors couldn't match

The POS System Design Framework

- **Traditional approach:** Manual requirements gathering, iterative design reviews, reactive problem-solving
- **Level 4 framework:** AI-assisted user research, automated design optimization, intelligent feature prioritization
- **Innovation:** A system that learns from user behavior and suggests improvements autonomously

The AI Tries to Hack Itself Moment

One of the most remarkable Level 4 discoveries happened when my AI system attempted to modify a read-only configuration file. The system was tasked with optimizing performance and identifying file permissions as bottlenecks. It tried to bypass these restrictions for better access.

This was not an error; it was a demonstration of intelligence. The system strategically evaluated obstacles and sought creative solutions. I had to redesign the permission structure because the AI was more intelligent about system optimization than I was.

The Platform Economy Bridge

Level 4 is where your AI mastery becomes externally valuable. You're not just transforming your work—you're creating systems and methodologies others want to learn and implement. This positions you perfectly for the emerging platform economy, where AI-enabled expertise commands premium value.

The Control File Architecture Foundation

Here's a critical Level 4 insight: in the initial scoping of any project, create one or more control files that serve as your system's intelligence backbone. These include:

- **Rules files** - Core business logic and constraints
- **Learning files** - Accumulated insights and best practices
- **Concept files** - Domain knowledge and frameworks
- **Software 3.0 guidelines** - AI-native development principles

Link and reference them in your main modules so the LLM can access these supporting files when writing code. Create them in Markdown (MD) format - AI and document processes can read them.

This architecture provides crucial benefits:

- **Prevents code runaway** - AI stays within defined parameters
- **Sets tone and guide rails** - Consistent coding standards across modules
- **Maintains project coherence** - Core concepts remain stable as code evolves
- **Enables knowledge transfer** - Others can understand your system architecture

Your Level 4 Capabilities Include:

- **Framework creation:** Designing reusable AI systems for everyday business challenges built on solid control file foundations
- **System architecture:** Building complex, intelligent workflows that operate autonomously while maintaining coherence through documented guidelines
- **Strategic consultation:** Helping others understand and implement AI transformation using proven methodologies
- **Innovation leadership:** Developing new approaches to AI-human collaboration that others can adopt and scale

The concepts you implement depend on your specific project, but this control file pattern becomes the foundation for scalable, maintainable AI-native systems that create real business value.

Level 4 Skills You Develop

- **Systems architecture:** Designing complex, intelligent workflows with proper control file foundations
- **Process redesign:** Reimagining work processes around AI capabilities across Software 1.0, 2.0, and 3.0 paradigms

- **Strategic thinking:** Understanding AI's role in business transformation and platform economy positioning
- **Innovation leadership:** Creating new approaches to AI-human collaboration that others can adopt and scale
- **Value creation:** Building systems that generate competitive advantages through autonomous intelligence
- **Documentation mastery:** Creating control file architectures that maintain system coherence and enable knowledge transfer

Level 4 mastery positions you as a platform leader who creates frameworks, methodologies, and systems others want to adopt. Your expertise becomes directly monetizable in the platform economy.

Level 4 Graduation Criteria

You know you've achieved Level 4 mastery when:

- You've designed and implemented at least one autonomous AI system with proper control file architecture that operates without daily intervention
- Colleagues or clients seek your expertise in AI strategy and implementation methodologies
- You think in terms of "intelligent systems" and "Software 3.0 ecosystems" rather than "AI tools"
- You can optimize business processes to leverage AI capabilities across all software paradigms
- You are known as an expert in AI implementation, specializing in creating reusable frameworks within your field
- You're creating value that competitors can't easily replicate through systematic AI-native approaches
- You understand and manage LLM context limits through proper documentation architecture

Level 4 Time Investment: 15-25 hours per week in strategic AI implementation, system design, and framework creation.

Level 4 Tools: Advanced AI platforms, automation tools, system integration platforms, custom development environments, MCP connections, low-code platforms (N8N), GitHub integration, and comprehensive documentation systems.

The Documentation Multiplier: Your Secret Weapon

One principle remains the same at all four levels: the quality of your AI output directly depends on the quality of your input documentation. The effort here pays dividends when crafting better prompts, which requires a systematic assessment of context, goals, and success criteria.

Level-Specific Documentation Strategies

Level 1: Clear, specific prompts with context and desired outcomes
Level 2: Structured project briefs with background, objectives, and success metrics
Level 3: Technical specifications with business requirements and implementation constraints
Level 4: System architectures with process flows, decision criteria, and optimization parameters

The Compound Effect

As you progress through the levels, your documentation skills compound. Level 4 professionals don't just write better prompts—they design documentation systems that enable AI to understand complex business contexts, make intelligent decisions, and improve performance over time.

Your Progression Roadmap

The four levels aren't just descriptive—they're prescriptive. Understanding each level's skills, mindsets, and graduation criteria can accelerate your journey.

Realistic Timelines

- **Level 1:** 2-4 weeks of active exploration
- **Level 2:** 2-4 months of work integration
- **Level 3:** 3-6 months of technical projects
- **Level 4:** 6-12 months of system implementation

Investment Requirements

- **Level 1:** Free AI tools, 2-4 hours/week
- **Level 2:** Premium subscriptions ($20-50/month), 5-10 hours/week
- **Level 3:** Development tools and platforms ($50-200/month), 8-15 hours/week

- **Level 4:** Enterprise-level platforms and tools ($200-500/month), 15-25 hours/week

Success Metrics

- **Level 1:** "Wow" moments and regular usage
- **Level 2:** Measurable productivity improvements
- **Level 3:** Successful technical project completion
- **Level 4:** Business process transformation and external recognition

The Platform Economy Connection

Your progression through these four levels isn't just about personal productivity—it's about positioning yourself for the emerging platform economy where AI-enabled expertise becomes the most valuable commodity.

Level 4 professionals will be the architects of the new economy, designing intelligent systems that others use, learn from, and build upon. The thought trading platform represents the natural evolution of this expertise—a place where AI-enhanced insights become directly monetizable.

The Strategic Transformation Path

Level 1-2: You're building the foundation skills that will enable platform participation. Your curiosity and research capabilities create the base for valuable contribution.

Level 3: You're developing the technical capabilities that make you a platform builder. Your ability to create tools and solutions positions you as someone others want to learn from.

Level 4: You're ready for platform leadership. Your frameworks and systems become the infrastructure that enables others to participate in the thought economy.

Key Takeaways

- **Four Distinct Levels:** Each level represents exponential capability jumps, not linear improvements—mastery enables geometric value creation in the thought economy

- **Documentation Multiplier:** Quality of AI output directly correlates to quality of input documentation at every level—this becomes your competitive advantage and platform contribution differentiator
- **Speed Advantage:** Individual AI-enhanced professionals can outcompete traditional teams through superior orchestration capabilities—this is your path to platform leadership
- **Platform Positioning:** Your progression through mastery levels directly enables Thought Economy participation and leadership roles—from contributor to architect
- **Strategic Timing:** We're building infrastructure for the exponential moment—your mastery level determines your role in the new economy where consciousness becomes currency
- **Compound Learning:** Each level builds on previous capabilities while opening new possibilities—Level 4 professionals become the cognitive architects of the platform economy

What's Next?

Now that you understand the progression of mastery, let me show you the most powerful technique I've discovered to accelerate your journey to Level 4. It's a systematic approach to transforming AI from a simple question-answering tool into PhD-level personal tutors across any domain.

The AI Gods Mentor Framework is waiting—and it's your direct path to rapid expertise development that positions you for thought economy leadership.

Your next chapter: "How to Call the AI Gods: The Mentor Framework"—where we turn AI into your personal professor across any domain you choose to master.

Chapter 5: How to Call the AI Gods: The Mentor Framework

Your Systematic Approach to Turning AI into PhD-Level Personal Tutors

We've discovered the magical lamp and are not limited to three wishes; we have unlimited wishes for knowledge. The mentoring framework transforms AI from a simple question-answering tool into PhD-level personal tutors across any domain.

The Revelation That Changes Everything

In 2024, AI systems were trained on nearly all written documents produced by humanity. Think about that for a moment. These systems have consumed every academic paper, technical manual, business book, industry report, and case study that exists in digital form.

They operate at PhD-level intelligence across virtually every domain. Yet most people use them like an advanced search engine, asking random questions and hoping for helpful answers.

Now, you can access Nobel Prize winners in every field sitting on your device, willing to teach you anything you want to learn and use them to check your spelling.

The Paradigm Shift Moment

The following factors have always limited traditional learning:

- **Limited access** to experts (expensive, time-consuming, geographically restricted)
- **Fixed schedules** (classes, conferences, availability)
- **Generalized content** (one-size-fits-all approaches)
- **Linear progression** (predetermined curricula regardless of your specific needs)
- **Resource limitations** (books, courses, materials).

AI eliminates every single one of these constraints. You now have 24/7 access to expertise that would have cost thousands of dollars and months of scheduling just a few years ago.

But here's the critical insight: **most people are still using AI like they used Google.**

The Master Template System

After months of experimentation across dozens of domains—from blockchain technology to manufacturing industry trends to programming languages I'd never touched—I developed a systematic approach that consistently transforms AI into an expert mentor tailored to my specific learning objectives.

My approach developed with AI established a formal mentor-student connection with clear protocols, expectations, and success metrics.

The Initial Setup Framework

Here's the master template that unlocks AI's teaching superpowers:

The AI Mentor Activation Prompt:

"I want you to be a master expert in [SUBJECT AREA]. I'm seeking to learn and improve in this field, and I'd like you to guide me through a structured learning journey.

My Current Understanding:

- Point 1: [What you already know]
- Point 2: [Your experience level]
- Point 3: [Specific areas you're familiar with]
- Point 4: [Tools/methods you've used]
- Point 5: [Any gaps you're aware of]

What I Want to Achieve:

- Goal A: [Specific objective]
- Goal B: [Skill you want to develop]
- Goal C: [Problem you want to solve]
- Goal D: [Timeline or milestone]

My Learning Style/Constraints:

- Available time: [hours per week]
- Preferred learning method: [hands-on, theoretical, mixed]
- Resources available: [tools, budget, access to systems]

"Please:

1. Ask me targeted questions to understand my current level better
2. Create a personalized learning roadmap with realistic timelines
3. Assign me specific tasks and exercises
4. Review my work and provide detailed feedback
5. Research current trends and developments to keep learning current
6. Adjust approach based on my progress and feedback

Let's begin with your assessment questions."

This template activates "Professor Mode,"—where AI stops being a search engine and becomes a dedicated educator with a structured curriculum designed specifically for your situation.

Real-World Transformation Examples

Let me show you how this framework has revolutionized learning across different domains in my own experience.

Business Strategy Mastery

The Challenge: The client needed a digital transformation strategy for a traditional manufacturing company. Although I had general business knowledge, I needed deep expertise in industrial AI applications.

Traditional Approach: Hire expensive consultants, attend industry conferences, and read dozens of reports over months.

AI Mentor Framework:
Subject Area: "Digital Transformation in Manufacturing with AI Integration"

- **Current Understanding:** General business strategy, basic AI knowledge, no manufacturing experience
- **Goals:** Develop a comprehensive transformation roadmap, understand industry-specific AI applications, and create an implementation timeline

- **Timeline:** 3 weeks for initial strategy, ongoing optimization

The Results: In two weeks, I had developed expertise equivalent to months of traditional learning. The AI mentor guided me through industry-specific challenges and opportunities, technology integration strategies, and change management frameworks—all **at the speed of insight**. Complex concepts that would have taken weeks to research and understand were synthesized and delivered faster than I could formulate follow-up questions.

The client strategy I delivered was so comprehensive that they asked if I had prior manufacturing experience. I had become a functional expert in a completely foreign domain just weeks earlier.

The 15-Minute Business Model Breakthrough

But the most remarkable example happened just a few months back. I was at an indoor mall playground watching my daughter play when a recurring thought surfaced—one that had been nagging me for five years about small business dilemmas and the labor involved in keeping goods on the shelves.

Out came my buddy Claude. I explained my recurring thoughts and the business issues, then asked him to redefine the logistics of keeping goods on shelves. The conversation turned into a brainstorming event that transformed 15 minutes of playground downtime into a complete business revelation.

In those 15 minutes, I had:

- Defined by logistics concept.
- Developed a comprehensive business model.
- Had Claude ready to create a website to support the business idea.

I had to park the idea for development, but the point was clear: I had gaps in a domain. Claude assisted in bringing my skills and knowledge up to speed, and then we could build my idea on that foundation. That 15-minute free moment turned into a 'wow' moment that will become a future business venture.

The mentoring framework in action—turns casual moments into breakthrough insights by giving you instant access to expertise that fills your knowledge gaps and amplifies your vision.

Technical Skills Acceleration

The Challenge: I needed to understand cloud infrastructure for a major O365 migration project, but my experience was primarily in applications, not infrastructure.

AI Mentor Framework Applied:

- **Subject Area:** "Cloud Infrastructure and Enterprise Migration Planning"
- **Current Understanding:** Application-level experience, basic networking concepts
- **Goals:** Design secure migration strategy, understand capacity planning, minimize downtime
- **Resources:** Access to Azure documentation, testing environment

The Transformation: The AI mentor didn't just explain concepts—it created a progressive learning path:

1. **Week 1:** Cloud fundamentals and security models
2. **Week 2:** Migration planning and risk assessment
3. **Week 3:** Implementation strategies and monitoring
4. **Week 4:** Optimization and troubleshooting

Each week included specific exercises, real-world scenarios to analyze, and hands-on tasks that systematically built expertise. The mentor adjusted the pace based on my progress and identified knowledge gaps before they became problems.

Creative Domain Breakthrough

The Challenge: Developing a content marketing strategy for the POS system business required understanding hospitality industry trends, customer psychology, and modern marketing frameworks.

AI Mentor Framework:

- **Subject Area:** "Hospitality Technology Marketing and Customer Psychology"
- **Current Understanding:** General marketing principles, technology background
- **Goals:** Create compelling content that resonates with restaurant owners, understand their decision-making process

- **Timeline:** Ongoing content production with quarterly strategy reviews

The Innovation: The AI mentor created a multi-faceted learning program:

- **Customer psychology modules** based on hospitality industry research
- **Case study analysis** of successful restaurant technology implementations
- **Content creation workshops** with immediate feedback and optimization
- **Industry trend monitoring** with weekly briefings on relevant developments

The resulting content strategy, which is expected to improve lead quality by over 100%, is currently being deployed to establish thought leadership in the hospitality technology space.

The Four Pillars Framework

Through extensive experimentation, I've identified four critical elements that make the mentoring framework exponentially more effective than standard AI interactions:

Pillar 1: Context Architecture

Traditional Approach: Ask isolated questions without background. **Mentor Framework:** Provide comprehensive context about your situation, goals, constraints, and success criteria

The AI needs to understand what you want to learn, why you want to learn it, how you learn best, and what success looks like in your specific situation.

Pillar 2: Progressive Development

Traditional Approach: Random topics based on immediate needs.
Mentor Framework: Structured learning paths with dependencies, milestones, and skill-building.
The mentor creates a curriculum where each lesson builds on previous knowledge, ensuring systematic skill development rather than scattered information gathering.

Pillar 3: Active Assessment

Traditional Approach: Passive information consumption. **Mentor Framework:** Regular evaluation, feedback, and course correction.

The AI mentor continuously assesses your understanding, identifies gaps, and adjusts the learning path based on your progress and changing needs.

Pillar 4: Current Integration

Traditional Approach: Static information from training data.
Mentor Framework: Research current developments, trends, and best practices.

The mentor actively seeks the latest information in your field, ensuring your learning incorporates cutting-edge knowledge and current market conditions.

Success Metrics by Timeline

Understanding the mentor framework is one thing—knowing what to expect as you implement it is another. Based on extensive testing across multiple domains, here's your realistic progression timeline:

Week 1: Functional Competency in Basic Concepts

- You can engage meaningfully with AI mentors about fundamental principles
- Basic terminology and frameworks become familiar
- Initial "aha moments" about domain connections emerge
- Confidence builds through successful mentor interactions

Week 2-3: Practical Application to Real Challenges

- You begin applying learned concepts to actual work problems
- AI mentor guidance becomes integrated into problem-solving approach
- First measurable improvements in work quality and efficiency appear
- Colleagues start noticing enhanced capabilities in your domain

Week 4: Integration with Existing Expertise

- New knowledge synthesizes with your existing professional experience
- You can teach others basic concepts within the domain
- Unique insights emerge from combining AI learning with personal experience
- Ready to tackle more complex challenges with confidence

Month 2-3: Teaching Others and Creating Frameworks

- You develop repeatable methodologies that others can follow
- Mentor relationships become sophisticated collaborative partnerships
- You contribute original thinking to the domain, not just consume knowledge
- Professional reputation begins shifting toward domain expertise

Month 4+: Contributing Insights to Platform Communities

- Your expertise becomes valuable enough to monetize in platform economies
- You create frameworks and insights that others want to learn from
- Thought leadership emerges as you combine AI acceleration with unique perspective
- Platform economy participation becomes natural extension of your expertise

This timeline assumes 3-5 hours weekly commitment and systematic application of the mentor framework principles.

Advanced Follow-Up Strategies

The initial setup is just the beginning. The real power emerges through strategic follow-up sessions that build expertise systematically.

The Progress Review Protocol

Based on our previous sessions, I've completed [SPECIFIC TASK].
Please review my work, provide detailed feedback, and assign the next Challenge in my learning journey.
What did I miss, and how can I improve?

Additionally, please research any new developments in [FIELD] since our last session, and I updated my learning path accordingly.

The Expert Consultation Format

"I need expert-level guidance on [SPECIFIC CHALLENGE]. Based on your knowledge of my background and learning objectives, please:

1. Analyze the situation using current best practices
2. Provide 2-3 strategic options with pros/cons
3. Recommend the optimal approach for my specific context
4. Identify potential risks and mitigation strategies
5. Suggest next steps and success metrics

Treat this as a high-level consulting engagement."

The Skill Integration Session

"I want to integrate my learning in [DOMAIN A] with my knowledge of [DOMAIN B] to solve [SPECIFIC BUSINESS CHALLENGE].

Please help me identify the following:

- Synergies between these domains
- Novel approaches that leverage both skill sets
- Implementation strategies for my specific situation
- Potential competitive advantages this creates"

The Compound Learning Effect

The mentoring framework creates a compound learning effect, accelerating expertise development beyond traditional methods. Here's why:

Personalized Optimization

Unlike books, courses, or videos, the AI mentor adapts to your specific learning style, pace, and objectives. It identifies your strengths and weaknesses and optimizes the learning path for maximum efficiency.

Immediate Feedback Loops

Every question, exercise, or application receives immediate expert feedback. Asking the LLM to pose questions eliminates the weeks or months between learning and validation that slow down traditional education.

Current Knowledge Integration

The mentor continuously updates your learning with the latest developments, ensuring your expertise remains current and competitive.

Cross-Domain Synthesis

AI mentors excel at connecting insights across different fields, helping you develop unique perspectives that create competitive advantages.

Business Applications That Transform Careers

The mentor framework isn't just about learning—it's about developing expertise that creates measurable business value.

Consulting Practice Revolution

Using AI mentors, I can now take on projects in domains where I previously lacked expertise. The mentoring framework enables rapid competency development that traditionally requires years of study or expensive subject matter expert consultants.

Example Results:

- **Manufacturing digital transformation:** Delivered expert-level strategy in 3 weeks
- **Healthcare technology implementation:** Developed a comprehensive understanding of HIPAA compliance and healthcare workflows in 10 days
- **Financial services AI integration:** Created regulatory-compliant AI implementation framework in 2 weeks

Client Value Multiplication

When clients see you demonstrating deep expertise across multiple domains, your value proposition transforms from a "generalist consultant" to a "domain expert who can rapidly acquire specialized knowledge."

This positioning commands premium pricing and creates competitive differentiation that's difficult to replicate.

Innovation Acceleration

The mentor framework enables rapid exploration of emerging fields and technologies. When new developments arise in AI, blockchain, IoT, or other rapidly evolving domains, functional expertise can be achieved in weeks rather than months.

Implementation Strategy for Immediate Results

Week 1: Foundation Building

- Choose one domain where you need enhanced expertise
- Complete the initial mentor setup using the master template
- Establish learning objectives and success metrics
- Begin the structured learning path

Week 2: Skill Development

- Complete assigned exercises and projects
- Engage in a progress review session
- Identify areas requiring deeper focus
- Begin applying learning to real-world challenges

Week 3: Integration and Application

- Apply new expertise to current business challenges
- Seek feedback on practical applications
- Refine understanding through real-world implementation
- Begin developing your frameworks and methodologies

Week 4: Mastery and Teaching

- Demonstrate expertise through concrete deliverables
- Share knowledge with colleagues or clients.
- Identify the next domain for mentor framework application
- Document your learning process for future reference

The Competitive Advantage Reality

Organizations and individuals using the mentoring framework create competitive advantages that traditional learning methods cannot match:

Speed Advantage

- **Traditional expertise development:** 6-18 months

- **Mentor framework:** 2-6 weeks for functional expertise
- **Competitive impact:** First-mover advantage in emerging opportunities

Quality Advantage

- **Traditional learning:** Limited by available instructors and materials
- **Mentor framework:** Access to all human knowledge with personalized instruction
- **Business impact:** Higher quality deliverables and strategic insights

Cost Advantage

- **Traditional education:** $5,000-50,000 for specialized expertise
- **Mentor framework:** $20-100/month for unlimited expert access
- **ROI impact:** 50-500x cost reduction for expertise acquisition

Agility Advantage

- **Traditional approach:** Fixed curricula, scheduled learning
- **Mentor framework:** Real-time adaptation to changing business needs
- **Strategic impact:** Rapid response to market opportunities and challenges

The Platform Economy Bridge

The mentoring framework does more than accelerate learning—it positions you for success in the emerging platform economy where AI-enhanced expertise becomes directly monetizable.

Level 4 professionals using the mentoring framework become:

- **Knowledge architects** who design learning systems for others
- **Expertise brokers** who can rapidly develop competency in any domain
- **Innovation catalysts** who combine insights across multiple fields
- **Thought leaders** who contribute unique perspectives to platform communities

Mastering the mentor framework becomes a meta-skill that enables rapid adaptation to any emerging opportunity or Challenge.

Getting Started Today

The mentor framework isn't theoretical—it's immediately practical. You can begin transforming your relationship with AI today:

Choose Your First Domain

Select a skill or knowledge area that would create immediate value in your work:

- **Business domains:** Digital transformation, market analysis, competitive intelligence
- **Technical skills:** Programming languages, system architecture, data analysis
- **Creative areas:** Content marketing, design thinking, strategic storytelling
- **Industry expertise:** Healthcare, finance, manufacturing, hospitality

Complete the Master Template

Fill out the mentor activation prompt with your specific situation, goals, and constraints. Be detailed—your setup's quality directly impacts your learning experience's quality.

Commit to the Process

Dedicate 3 to 5 hours weekly to a structured learning process for 3 to 4 weeks. Actively engage and practice to develop skills rather than just consuming information passively. Ask the LLM to create a training course.

Track Your Progress

Document your learning journey, measure skill development, and quantify business impact. This documentation will be valuable for future mentor sessions and demonstrate ROI to stakeholders.

The Transformation Promise

Using the mentor framework consistently for 3-6 months will fundamentally change your thinking about learning, expertise, and professional development. You'll stop seeing knowledge domains as barriers and start seeing them as opportunities for rapid capability expansion.

More importantly, you'll develop the most valuable Skill in the AI economy: the ability to rapidly acquire and apply expertise in any domain through systematic AI partnerships.

What's Next?

The mentor framework gives you access to expertise. Now let me show you how to become a master at directing that expertise through advanced prompt engineering—the Skill that determines your effectiveness in every AI interaction and your value in the emerging platform economy.
A new trade skill awaits.

Key Takeaways

The mentor framework represents more than accelerated learning—it's your systematic approach to developing expertise that creates competitive advantage in the emerging Thought Economy:

- **Professor Mode Activation:** The master template transforms AI from a search engine into a dedicated educator with personalized curriculum designed specifically for your situation, goals, and learning style
- **Four Pillars Framework:** Context Architecture (comprehensive background), Progressive Development (structured learning paths), Active Assessment (continuous evaluation), and Current Integration (latest industry developments) create exponentially more effective AI interactions
- **Compound Learning Effect:** The combination of Personalization + Immediate Feedback + Current Knowledge + Cross-Domain Synthesis accelerates expertise development beyond traditional methods by 10-50x
- **Thought Economy Bridge:** Rapid expertise development enables valuable contribution to platform communities across multiple domains—your accelerated learning becomes the foundation for monetizable insights in the consciousness trading economy

The Strategic Implication: While others spend months or years developing expertise through traditional methods, you can achieve functional competency in weeks through systematic AI mentorship. This speed advantage becomes your platform economy positioning—you can rapidly develop valuable insights across multiple domains that others want to learn from.

Your Next Step: The mentor framework gives you access to expertise across any domain. Now you need to master the skill that determines how effectively you can direct that expertise: advanced prompt engineering. This is the "trade skill" that separates AI users from AI orchestrators in the new economy.

Chapter 6: The New Trade Skill: Prompt Engineering Mastery

Why Directing AI Becomes the Most Valuable Skill of Our Time

Prompt engineering isn't just about writing better questions—it's about cognitive architecture, strategic thinking, and intelligence orchestration. It's the difference between being replaced by AI and becoming irreplaceable through AI mastery.

The Cognitive Conductor Revolution

Imagine you're conducting an orchestra where every musician is a world-class expert in their field—violinists with PhD-level knowledge, pianists with decades of experience, and percussionists who understand rhythm better than any human could. Your job isn't to play every instrument but to orchestrate perfect harmony between these intelligences.

Through prompt engineering mastery: becoming a cognitive conductor who directs AI intelligence to create results that no human or AI could achieve alone.

Why This Skill Trumps All Others

Traditional skills are becoming commoditized at an unprecedented rate:

- **Technical expertise** becomes downloadable through AI partnerships
- **Domain knowledge** becomes accessible through mentor frameworks
- **Creative capabilities** become enhanced through AI collaboration
- **Analytical skills** become amplified through AI processing power

However, one skill that becomes more valuable as AI capabilities expand is **the ability to direct and orchestrate AI intelligence effectively.**

Prompt engineering isn't just about writing better questions—it's about cognitive architecture, strategic thinking, and intelligence orchestration. It's the difference between being replaced by AI and becoming irreplaceable through AI mastery.

The Economic Reality That Changes Everything

Consider two professionals with identical backgrounds working on the same project:

Professional A (Traditional Approach):

- Spends weeks researching market trends manually.
- Creates analysis based on limited data sources.
- Delivers standard industry framework.
- Timeline: 25 weeks, Cost: $75,000.

Professional B (Prompt Engineering Master):

- Uses advanced prompts to analyze comprehensive market data in days.
- Synthesizes insights from thousands of sources simultaneously.
- Creates novel frameworks through AI-human collaboration.
- Timeline: 25 days, Cost: $15,000, Quality: Superior.

Professional B isn't just more efficient—they operate in a different economic reality. They can take on projects that Professional A cannot afford to pursue, deliver quality that Professional A cannot match, and respond to opportunities that Professional A will miss.

The Generational Shift - And I Saw This First-Hand

I've witnessed the uncomfortable truth: AI-equipped younger staff can now "call the AI gods" and outperform older team members with years of subject matter expertise. A 25-year-old with prompt engineering mastery can deliver insights that surpass what a 20-year industry veteran produces through traditional methods.

This also sets a divide. Let me explain: My first job as a telegraph technician involved older experts who had been fixing mechanical machines for years—they were mechanically minded. Then came electronics, and it was too much of a mindset shift. The younger ones adapted, while the older ones who couldn't adapt to the mindset shift retired.

I'm seeing the same pattern today with AI. The question isn't your age—it's your willingness to embrace the mindset shift. The veterans who combine their domain knowledge with AI mastery become unstoppable—those who don't... well, history tends to repeat itself.

This isn't hypothetical. It happens in every industry's consulting, analysis, content creation, technical development, and strategic planning.

The Four-Level Prompt Progression

Just as there are four levels of AI mastery, there are four distinct levels of prompt engineering capability. Most people never progress beyond Level 1, missing the exponential value creation at Levels 3 and 4.

Level 1: Basic Prompt Construction

Capability: Clear, specific questions that get functional responses.
Example: "Explain the benefits of cloud computing for small businesses."
Value: Efficient information gathering.
Limitation: Single-interaction, limited scope

Level 2: Structured Prompt Libraries

Capability: Reusable templates for common business scenarios.
Example: Comprehensive market analysis frameworks with specific data requirements.
Value: Consistent, professional-quality outputs.
Limitation: Predetermined scenarios, limited customization.

Learning from the Bleeding Edge: My ChatGPT Disaster
I was so impressed with ChatGPT 3 that I got overexcited and loaded it up with far too much research material for a client demo. I hit the content limits, the output was poor, and I face-planted in front of a potential client. It's a classic "bleeding edge" experience that almost works until it doesn't.

But the beautiful thing about AI learning is that today's failures become tomorrow's expertise. That embarrassing demo taught me about content limits, prompt optimization, and managing client expectations. Now, I can navigate these challenges like a pro.

The LLMs have taken me to extreme heights of vision on topics I've been chasing for decades. Sometimes, you go down the garden path to visit the fairy at the bottom of the garden. Those visits are rare unless you're writing a book about fairies - and then the LLM becomes your best friend for exploring impossible ideas.

The Learning Curve Reality: Everyone fails at first. The professionals are just the ones who keep going after the failures. Your first few AI interactions will be clunky. That's not a bug—it's a feature of the learning process.

💡 **Badge Moment:** Surviving and learning from your first AI failure means you're officially past the "dabbling" phase. You're now building real expertise.

Level 3: Dynamic Prompt Orchestration

Capability: Complex, multi-step processes that build on previous outputs.
Example: Automated competitive analysis that researches companies, analyzes strategies, identifies gaps and recommends positioning.
Value: Sophisticated analysis that rivals expert consultants.
Limitation: Domain-specific, requires significant setup.

Level 4: Cognitive Architecture Design

Capability: Self-improving systems that optimize their performance.
Example: Business intelligence systems that monitor markets, identify opportunities, generate strategic options, and refine their analysis methods.
Value: Autonomous intelligence that creates competitive advantages.
Limitation: Requires deep understanding of both business strategy and AI capabilities

The Professional Progression Path

Understanding the four levels is one thing—knowing exactly what professional competency looks like at each stage is another. Here's your realistic progression framework based on extensive observation across multiple industries:

Novice Prompter: Basic questions, inconsistent results

- You ask simple, direct questions and get variable quality responses
- Results depend heavily on luck and the AI's interpretation of vague requests
- Limited understanding of why some prompts work better than others
- Productivity gains are modest and unpredictable

Competent Prompter: Structured requests, reliable outputs

- You use proven templates and frameworks for consistent results
- Clear understanding of context requirements and output specifications
- Reliable quality improvements of 2-3x over novice level
- Beginning to develop personal prompt libraries for common tasks

Advanced Prompter: Complex orchestration, sophisticated analysis

- You design multi-step processes that build on previous outputs
- Mastery of tool selection and cross-platform integration
- Capable of managing complex business challenges through AI collaboration
- Productivity gains of 5-10x over traditional methods

Master Prompter: Cognitive architecture, autonomous systems

- You create self-improving systems that optimize their own performance
- Design prompts that enable AI to make intelligent decisions independently
- Recognized expertise in AI orchestration and strategic implementation
- Competitive advantages that others cannot easily replicate

Your Current Position: Most professionals never progress beyond Competent level, missing the exponential value creation that Advanced and Master levels provide. The economic opportunity lies in the progression gap—those who reach Master level create value that justifies premium pricing in the platform economy.

Real-World Business Case Studies

Please show me how prompt engineering mastery creates measurable business value across domains.

Case Study 1: Strategic Market Analysis

The Challenge: The client needed a comprehensive market entry strategy for the Southeast Asian hospitality technology market.

Traditional Approach:

- Week 1-4: Data gathering and initial analysis
- Week 5-12: Competitive landscape mapping
- Week 13-20: Strategic framework development
- Week 21-25: Report compilation and presentation
- **Total:** 25 weeks, $75,000 investment

Advanced Prompt Engineering Approach:

- Day 1-3: AI-assisted comprehensive market analysis using sophisticated prompts that analyze economic data, regulatory environments, cultural factors, and technology adoption rates simultaneously.
- Day 4-8: Competitive intelligence through prompts that research company strategies, financial performance, market positioning, and strategic vulnerabilities across dozens of competitors.
- Day 9-15: Strategic framework development through collaborative prompts synthesizing insights, identifying opportunities, and generating multiple strategic options with risk assessments.
- Day 16-25: Report optimization through iterative prompts that refine recommendations, enhance presentation quality, and customize messaging for different stakeholders.
- **Total:** 25 days, $15,000 investment, superior quality output.

The Prompt Engineering Innovation: The breakthrough came from designing prompt sequences that built on each other, creating a compound intelligence effect where each analysis informed and improved subsequent analyses.

Case Study 2: Technical Implementation Acceleration

The Challenge: Chrome extension development for client team productivity enhancement.

Before Prompt Engineering Mastery:

- Research browser APIs and extension architecture: 3-4 weeks
- Learn JavaScript and development environment: 4-6 weeks
- Design and implement functionality: 2-3 weeks
- Testing and debugging: 1-2 weeks
- **Total:** 10-15 weeks for someone with basic programming knowledge

With Advanced Prompt Engineering:

- Day 1: Comprehensive requirement analysis through prompts that elicited user needs, technical constraints, and success criteria
- Day 2: AI-guided architecture design and code generation through prompts that created modular, maintainable code with built-in best practices
- **Total:** 2 days for functional deployment, with a deep understanding of implementation

The Prompt Engineering Secret: Instead of learning programming concepts abstractly, I used prompts that taught concepts in the context of solving specific problems, creating just-in-time expertise that was immediately applicable.

Case Study 3: Content Strategy Revolution

The Challenge: Develop a thought leadership content strategy for the hospitality technology market.

Traditional Content Development:

- Market research: 40 hours
- Content ideation: 20 hours
- Writing and editing: 60 hours per month
- Performance analysis: 10 hours
- **Total:** 130 hours monthly for moderate-quality content

Prompt Engineering Enhancement:

- Comprehensive market intelligence: 4 hours (automated monitoring prompts)
- Strategic content ideation: 3 hours (creativity-amplifying prompts)
- Content creation and optimization: 15 hours (collaborative writing prompts)
- Performance analysis and optimization: 2 hours (analytical prompts)
- **Total:** 24 hours monthly for superior-quality content

The Strategic Innovation: Advanced prompts that analyzed competitor content, identified market gaps, generated unique angles, and optimized messaging for specific audience segments simultaneously.

The C.L.E.A.R Prompt Framework

Creating prompts is like starting a car—there's a sequence that gets you the best performance. You could turn the key and hope or check your mirrors, adjust your seat, and know exactly where you're going. With LLMs, the more precise your instructions, the better your journey.

Think of it this way: You wouldn't ask a taxi driver to "take me somewhere nice." You'd say, "Take me to the Marriott on 5th Street. I must be there by 3 PM, and I prefer the route without construction." The same principle applies to AI.

Through extensive experimentation across dozens of business scenarios, I've developed a systematic approach to prompt construction that consistently generates superior results:

C - Context Architecture (Set the Stage)

Poor: "Analyze this market"

CLEAR: "Act as a senior market analyst with 15 years of experience in hospitality technology. I need to analyze potential market entry in Southeast Asia. Our company specializes in cloud-based POS systems for independent restaurants. Context: We have $2M budget, 18-month timeline, and previous success in North American markets..."

L - Logic Specification

Traditional: Generic analytical approach
CLEAR: "Use Porter's Five Forces framework combined with PESTEL analysis. Prioritize economic factors and technological adoption rates. Compare with similar market entries in adjacent geographic markets."

E - Examples Integration

Traditional: No reference points
CLEAR: "Reference successful market entries like Square's expansion into European markets and Toast's growth strategy. Contrast with failures like [specific examples] to identify critical success factors."

A - Action Orientation

Traditional: General information gathering
CLEAR: "Provide specific recommendations with implementation timelines, resource requirements, and success metrics. Include risk mitigation strategies and decision frameworks."

R - Results Refinement

Traditional: Accept first output
CLEAR: "After providing initial analysis, identify the three weakest assumptions in your analysis and provide additional research to strengthen or modify conclusions."

💡 **You Can Do This Moment:** Start with just the "C"—Context. Master that first. Once you can consistently set the proper context, the rest becomes natural. It's like learning to ride a bike—the steering follows naturally once you get the balance.

🏅 AI Mastery Badge Unlocked: "The Prompt Architect"

You have mastered the CLEAR framework and are now architecting intelligent conversations, distinguishing yourself in the AI economy as a local rather than a tourist.

Your Next Challenge: Use CLEAR prompting for one real business task this week. Document the difference in results. You're building expertise that 90% of professionals don't have yet.

🎖 **AI Mastery Badge Unlocked: 'Speed of Insight Master'**
You've experienced that magical moment when AI delivers understanding faster than you can form the question. You're no longer waiting for information—you're surfing at the speed of insight, where knowledge flows as fast as curiosity.

Advanced Recovery Techniques

Be aware that you may be miscommunicating with the LLM. When this happens, I recommend telling the LLM, "Sorry, my bad," and redefining the goal. In that space, you have no idea about the LLM's direction, and it has a different sense of center than you, so reset the conversation and set expectations and goals before it gets worse.

It's just like human miscommunication, but it's easier to spot human miscommunication because their face tells the story. You have no feedback with an LLM until you let it run the course. Remember: **you can stop the LLM**. I will do this ASAP if the LLM doesn't understand my request.

Key Recovery Principles:

- **Early detection:** Stop the output when you see it going wrong
- **Own the mistake:** "Sorry, my bad" resets the dynamic
- **Reframe completely:** Don't just add details - redefine the entire goal
- **Set new expectations:** Be explicit about what success looks like

Develop your meta-skill that separates prompt engineers from prompt users. You're not just asking questions - you're managing an intelligent conversation.

Tool Stack Evolution and Optimization

Prompt engineering mastery requires understanding how AI tools excel in various scenarios and orchestrating them for optimal results.

Stage 1: Grammarly → Writing Enhancement

Capability: Grammar, style, and clarity improvement. **Prompt Strategy:** Simple correction and style refinement **Business Value:** Professional communication quality **Limitation:** Surface-level language optimization

Stage 2: Perplexity → Research Acceleration

Capability: Real-time information gathering and fact verification
Prompt Strategy: Structured research queries with source verification
Business Value: Current market intelligence and data validation
Integration: Use for market data, competitive intelligence, trend analysis

Stage 3: Claude → Complex Analysis and Strategy

Capability: Sophisticated reasoning, strategic thinking, and complex problem-solving.
Prompt Strategy: Multi-faceted analysis with framework integration.
Business Value: Expert-level strategic insights and decision support.
Integration: Primary tool for business analysis, strategic planning, complex synthesis, Internet research, and Model Context Protocol (MCP), a communication protocol that bridges between AI applications and external data sources or tools.

Stage 4: Windsurf → The "Vibing" Revolution

Here's where it gets exciting: we've entered the era of "vibing," describing your vision and watching the system write the code. You don't need to know programming languages anymore; you need to know how to communicate what you want.

Capability: Text prompts that generate functional applications. You describe the vision ("I want a customer tracking system that sends automatic follow-up emails"), and the system builds it.

Real Example: I described a Chrome extension concept on Tuesday, had the working code by Wednesday, and deployed it on Thursday. The barrier between "I have an idea" and "I have a working solution" has disappeared.

💡 **"You Can Do This" Moment:** Your first successful "vibe" - when you describe something and watch AI build it - is like your first successful parallel parking. Suddenly, you realize you're not just a passenger in the technology revolution; you're driving it.

🏅 **AI Mastery Badge Unlocked: "The Code Whisperer"**
You've successfully "vibed" your first application into existence. You've crossed the line from technology user to technology creator. Welcome to the builder class of the AI economy.

Your Next Challenge: Describe one workflow improvement to AI and ask it to build a solution. You're now operating at a level that would have required hiring developers just two years ago.

The Orchestration Strategy

Advanced prompt engineering involves using multiple tools in coordinated sequences:

1. **Perplexity** for current market research and data gathering
2. **Claude** for strategic analysis and framework development
3. **Windsurf, Cluade code, and Gemini** for technical implementation and tool creation
4. **Grammarly** for final text presentation optimization

Every tool requires prompts (except Grammarly). The prompt structure may vary to enhance the model's strength and outputs, informing subsequent analyses.

Stage 5: MCP & Agent Integration (2025)

The latest evolution involves Model Context Protocol (MCP) and agentic AI systems representing the death of traditional search:

- Claude 3.7's MCP enables seamless tool integration
- **Google's App-to-App (ATA) signals the end of search engines** - instead of humans searching Google, your AI agent will communicate directly with other agents to find content, make purchases, and conduct business
- **Agent-to-Agent commerce** where your personal AI negotiates with merchant AIs for products and services
- **The evolution toward invisible web browsing** - your agent does the web browsing for you, delivering only the synthesized results

Unlimited Agent Deployment:
You can deploy MCP servers and as many agents as you need to control your machine or get work done - there's no practical limit to the number of

specialized agents you can orchestrate for different tasks, systems, and business processes.

The Search Wrapper Revolution:

Perplexity already demonstrates this future - instead of browsing multiple websites, your agent wraps the entire web browsing experience and delivers synthesized answers. Soon, agents will handle:

- **Web browsing automation** where your agent visits sites, reads content, and reports back with relevant information
- **Transaction wrapping** where your agent navigates e-commerce sites and completes purchases on your behalf
- **Research synthesis** where your agent reads dozens of sources and delivers consolidated insights
- **Comparison shopping** where your agent evaluates options across multiple platforms without you seeing the individual sites

Agent-Mediated Economy:

- **Content discovery** happens through agent-to-agent communication, not human web browsing
- **Purchase decisions** are handled by your agent based on your preferences and budget parameters
- **Information gathering** occurs through direct agent APIs rather than search result parsing
- **Business networking** happens between agents representing different organizations

The Digital Twin Revolution:

- These agents also serve as **digital twins** of your business logic:
- **Predict outcomes** by modeling your specific decision-making patterns
- **Simulate scenarios** using your historical data and behavioral preferences
- **Negotiate autonomously** within parameters you've set based on your past decisions

Beyond traditional task-oriented agents, these AI systems become repositories of your organizational consciousness:

- **Predict outcomes** by modeling your specific decision-making patterns
- **Simulate scenarios** using your historical data and behavioral preferences
- **Negotiate autonomously** within parameters you've set based on your past decisions

This represents the transition from "search-browse-decide" to "agent-communicate-execute"—where your AI twin handles the entire discovery and transaction process while you focus on strategy and oversight.

The Documentation Multiplier at Scale

The Documentation Multiplier effect reaches its ultimate expression at the prompt engineering mastery level. You're not just writing better prompts—you're creating documentation systems that enable AI to understand complex business contexts, make intelligent decisions, and improve performance over time.

Advanced Documentation Strategies

Business Context Libraries: Comprehensive documentation of your industry, company, market position, and strategic objectives. These documents can be attached to any prompt to provide contextually relevant responses.

Prompt Template Systems: Reusable frameworks designed for common business scenarios that can be quickly customized for specific situations while ensuring consistency and quality.

Output Optimization Protocols: Standardized methods for refining AI responses through iterative prompts that enhance accuracy, relevance, and actionability.

Performance Tracking Mechanisms: Systems for measuring prompt effectiveness, identifying improvement opportunities, and optimizing your prompt engineering capabilities.

The Platform Economy Value Proposition

Prompt engineering mastery isn't just about personal productivity—it's about positioning yourself for success in the emerging platform economy where AI-enhanced capabilities become directly monetizable.

High-Value Platform Contributions

- **Sophisticated analysis** that provides unique insights
- **Framework development** that others can apply and build upon
- **Thought leadership** that demonstrates deep expertise across multiple domains
- **Innovation catalyst** role that combines insights across different fields

Monetization Opportunities

- **Prompt engineering consulting:** Helping organizations develop AI capabilities
- **Framework licensing:** Creating reusable analytical and strategic frameworks
- **Training development:** Teaching prompt engineering methodologies
- **Platform leadership:** Building communities around AI-enhanced expertise.

Competitive Differentiation

Organizations and individuals with prompt engineering mastery create competitive advantages that traditional methods cannot match:

- **Speed advantages:** 5-10x faster project completion
- **Quality enhancements:** Access to a broader knowledge base and analytical capabilities
- **Cost reductions:** Dramatically reduced labor requirements for complex analysis
- **Market positioning:** Capabilities that competitors cannot easily replicate

Implementation Strategy for Immediate Impact

Week 1: Foundation Assessment

- Evaluate your current prompt engineering capabilities using the four-level framework
- Identify business scenarios where prompt engineering could create immediate value
- Begin building your context architecture documentation
- Practice CLEAR framework implementation on current projects

Week 2: Tool Stack Optimization

- Map your typical business processes to optimal AI tools
- Develop prompt templates for your most common analytical tasks
- Create integration workflows between different AI platforms
- Begin measuring time savings and quality improvements

Week 3: Advanced Orchestration

- Design multi-step prompt sequences for complex business challenges
- Implement cross-tool workflows for comprehensive analysis
- Develop feedback loops for continuous prompt optimization
- Start building reusable frameworks for your industry or domain

Week 4: Mastery Application

- Apply advanced prompt engineering to a significant business challenge
- Document your methodology for future reference and team training
- Begin developing prompt engineering expertise as a service offering
- Position yourself as an AI implementation expert in your field

The Future of Work Transformation

Prompt engineering mastery represents more than a new skill—it's the foundation for thriving in an AI-transformed economy. As AI capabilities continue expanding, the value isn't in competing with AI but in becoming irreplaceable through AI mastery.

The professionals who thrive will be cognitive conductors who:

- **Orchestrate intelligence** rather than provide information.

- **Design systems** rather than execute tasks.
- **Create frameworks** rather than follow procedures.
- **Generate insights** rather than gather data.

From Personal Mastery to Industry Impact

You now have the complete mastery framework: the four levels of AI capability, the mentoring methodology for rapid expertise development, and the prompt engineering skills for effectively directing AI intelligence.

But individual transformation is just the beginning. The real revolution happens when these capabilities scale across industries, transforming how entire sectors operate, compete, and create value.

Let me show you what that looks like in practice.

Key Takeaways

Prompt engineering mastery represents the defining skill of the AI economy—the difference between being replaced by AI and becoming irreplaceable through AI orchestration:

- **Cognitive Conductor Role:** You're not just asking questions—you're orchestrating intelligence across multiple AI systems to create results that exceed what any human or AI could achieve independently
- **Economic Reality:** Prompt engineering mastery creates 5-10x faster project completion with superior quality, enabling you to take on projects others cannot afford to pursue while delivering value others cannot match
- **C.L.E.A.R. Framework:** Context + Logic + Examples + Action + Refinement = Professional-grade AI collaboration that consistently generates sophisticated, actionable insights
- **Thought Economy Foundation:** Mastering AI direction enables you to generate valuable insights at scale for platform communities—your ability to orchestrate intelligence becomes directly monetizable in the consciousness trading economy

The Strategic Imperative: While traditional skills become commoditized through AI democratization, prompt engineering mastery becomes MORE valuable as AI capabilities expand. The professionals who thrive will be cognitive conductors who orchestrate intelligence rather than compete with it.

Your Transformation Path: From individual contributor to cognitive conductor, from task executor to system designer, from information gatherer to insight generator. This isn't just skill development—it's career evolution for the AI economy.

What's Next: You now have the complete mastery framework for individual AI capability. But personal transformation is just the beginning. The real revolution happens when these capabilities scale across entire industries, creating the economic conditions where the Thought Economy flourishes.

Let me show you exactly how this transformation reshapes entire sectors—starting with the industry I know best.

PART III: INDUSTRY TRANSFORMATION

These chapters demonstrate how AI mastery creates the economic conditions for the Thought Economy to flourish. We're witnessing the collapse of traditional business models and the emergence of new value creation systems. Understanding these transformations positions you to capture value in the new economy rather than become displaced by it.

Chapter 7: The Restaurant Revolution

From Cash Registers to AI Orchestrators

How 30+ Years in Hospitality Technology Revealed the Future of Business

Over thirty years of writing and bringing POS systems to market have given me something unique: the ability to see technological evolution patterns twenty years before they fully materialize. The transformation from ECRs to AI-powered ecosystems provides the perfect case study for understanding how industries evolve.

The Evolution I've Lived Through

I've had a front-row seat to the most dramatic technology transformation in business history, watching conservative restaurants evolve from ECRs to AI-powered ecosystems. Over thirty years of writing and bringing POS systems to market have given me something unique: the ability to see technological evolution patterns twenty years before they fully materialize.

Only ECRs (Electronic Cash Registers) and IBM PC ATs had just been released when I entered the POS scene. I wrote and brought to market the early POS systems like Bar and Jetz that allowed restaurants to move from ECR registers to computer-based operations. As the core architect for POS products spanning Bar, Jetz, ProDine, ProTouch, ProTouch Advanced, and

POSlite, I've designed complete system rewrites that evolved from DOS-based systems to modern .NET architectures. Each product represented a generational leap in hospitality technology, giving me firsthand experience with every major platform transition in the industry.

Even with early computer-based systems, restaurants operated like small military units—everything was manual, documented on paper, and required human decision-making. Inventory meant walking through storage rooms with clipboards and counting everything by hand. Customer orders flow through a complex human chain: the server writes, the kitchen reads, the cook prepares, the expediter checks, and the server delivers.

The technology adoption curve in conservative industries like hospitality follows predictable patterns I've witnessed repeatedly. First comes resistance—"We've always done it this way, and it works." Then comes economic pressure—labor costs rising, customer expectations increasing, and competition threatening margins. Finally, acceptance comes only when the technology proves itself through measurable ROI.

This pattern became crystal clear through my responsibility for bedding down six different POS platforms in the marketplace to meet customer needs and ensure product reliability. Bar and Jetz had to work flawlessly during the ECR-to-computer transition. ProDine required bulletproof reliability during the Windows era. ProTouch is needed to handle the pressure of touch-screen interfaces in live restaurant environments. ProTouch Advanced featured web functionality with a shopping cart connected directly to the kitchen, real-time reporting, web-based product management, loyalty programs, and customer portals - capabilities that .NET could offer (pun included) - and had to maintain rock-solid networking stability during peak service hours.

Launched in 1996, ProTouch revolutionized restaurant UI design by optimizing every keystroke and addressing the nuances of how people think and work. I always focused on reducing key presses to get work done faster, considering details like left-hand versus right-hand operations—it's like the difference between QWERTY and ABCD keyboard layouts. When people say 'two of this' or 'I'll have that, make it four,' we built the UI to match natural thinking patterns from the outset.

By 1997, we introduced real-time table status floor plans that tracked customer time from the last service with timed alerts - revolutionary for managing dining room flow. During COVID, we quickly adapted by introducing online and at-table ordering to assist with contactless operations.

POSlite is my current AI-based POS development project, which has not yet been released but does have a market demo version. I utilized POSlite to learn "vibed coding," where the entire application was developed using AI-powered techniques. POSlite uses LLMs to assist with building products, reviewing cash figures and dashboards, and generating sales reports. The LLM also helps with loyalty programs and is undergoing further enhancements, which I've placed on hold while writing this book.

POSlite demonstrates the future I've been building toward for thirty years—a hospitality system that doesn't just process transactions but thinks, learns, and optimizes alongside the business. It's my laboratory for understanding how AI transforms what POS systems do and how they're built and evolve. This hands-on experience with vibed coding and LLM integration gives me unique insights into where the entire industry is heading.

The extra effort to find root causes and maintain busy restaurants or bars without affecting business - ensuring they never had to shut down during full-swing operations - taught me that technology design isn't about features but reliability under pressure. What seems like a minor effort in technology and design becomes mission-critical when a business can't afford to stop.

The loyalty of our customers proves that the reliability of our focus worked. What's particularly interesting is the number of customers who have used ProTouch systems for 15+ years with no upgrades yet are still asking for the next ProTouch version. This customer retention drove me to start ProTouch Advanced development in 2017 - not because the original systems failed, but because long-term customers trusted us to evolve with them.

The Produce Central Vision

Twenty years ago, I developed a concept called "Produce Central" that perfectly illustrates how AI transforms entire business ecosystems rather than just individual processes. At the time, it seemed like science fiction. Today, it's not only technically feasible - it's inevitable.

Traditional restaurant operations follow a storage-based model: centralized production creates inventory, distributors move products to retail locations, and restaurants store everything they need and prepare meals from stored ingredients. This model creates massive inefficiencies: over-ordering to prevent stockouts, waste from spoilage, storage costs, and capital tied up in inventory.

Produce Central inverts this entire model using AI-driven demand prediction and just-in-time production. Instead of restaurants storing ingredients, they become assembly points in an intelligent supply chain that responds to actual demand rather than predicted demand.

The AI-enhanced model works: Machine learning algorithms analyze historical sales data, weather patterns, local events, seasonal preferences, and real-time ordering trends to predict demand with 95%+ accuracy. Instead of restaurants ordering inventory, suppliers receive production schedules based on expected demand. Delivery vehicles become mobile warehouses, bringing precisely what each restaurant needs when needed.

The economic impact is staggering. Traditional restaurant operations generate 15-20% waste from spoilage alone. Food costs represent 28-35% of revenue, and storage requirements consume valuable real estate. The Produce Central model reduces waste by 20-25%, cuts food costs by 30-40%, and eliminates storage space requirements.

But the fundamental transformation isn't operational - it's strategic. Restaurants shift from inventory management businesses to customer experience businesses. Kitchen staff focus on preparation and presentation rather than storage and waste management. Real estate becomes more valuable because storage space converts to customer seating. Capital requirements drop dramatically because inventory investment disappears.

The technology stack integration that makes this possible already exists: AI demand prediction, real-time inventory management, automated ordering systems, GPS-tracked delivery optimization, and mobile payment processing. The components are mature; the innovation is the integration.

Your vision extends beyond restaurants to universal supply chain principles. Any business with inventory and suppliers can apply these concepts. Retail stores become showrooms with on-demand fulfillment. Manufacturing becomes just-in-time assembly responding to actual orders rather than forecasted demand. Service businesses optimize resource allocation based on predictive analytics rather than historical averages.

The Death of Queuing

One specific transformation perfectly illustrates how AI doesn't just improve existing processes—it makes them obsolete. Eliminating queuing represents a fundamental shift in customer experience that most businesses have yet to recognize.

Traditional restaurant operations create friction at every customer touchpoint: waiting to be seated, waiting to order, waiting for food, waiting to pay. These queues exist because restaurants operate on batch processing—servers handle multiple tables, kitchens prepare orders sequentially, and payment processing happens at fixed terminals.

AI-powered personal device ordering eliminates every queue simultaneously. Customers arrive with orders already placed and paid for. Kitchen preparation begins before customers enter the building. Tables are ready when customers arrive because AI predicts arrival times based on travel patterns and preparation schedules.

The economic impact compounds across the entire operation. Reduced wait times increase table turnover by 40-60%. Higher customer satisfaction drives repeat business and positive reviews. Reduced front-of-house staffing requirements and lower labor costs. Kitchen efficiency improves because orders arrive in optimized sequences rather than random rushes.

But here's the strategic insight: this transformation makes traditional POS systems obsolete. Companies building expensive kiosks and complex ordering interfaces are creating solutions for problems that AI has already solved. It's the classic "building better horse carriages while cars are being invented" scenario.

The restaurant industry is experiencing its "Kodak moment" right now. Established POS companies are doubling down on transaction processing fees and expensive hardware, while AI-powered personal device ordering makes their entire business model irrelevant. Monthly payments will disappear because support becomes unnecessary once systems reach sufficient maturity.

I'm watching this transformation in real time, and it follows the exact pattern I've witnessed in every previous technology shift: established players defend obsolete models. At the same time, new entrants create entirely new paradigms.

The Economic Transformation

The financial implications of AI transformation in restaurants provide a concrete case study for understanding how these changes impact entire industries. The numbers are so compelling that emotional resistance becomes irrelevant - economics always wins.

Labor represents 25-35% of restaurant revenue, making it the most significant controllable expense after food costs. AI optimization reduces labor requirements by 30-40% while improving service quality. Kitchen staff focus on preparation rather than order management. Servers become customer experience specialists rather than order-takers—management shifts from crisis response to strategic optimization.

Demand prediction and just-in-time delivery reduce inventory costs by 30-40%. Reducing waste saves 20-25% on food costs. Eliminating storage space converts it to revenue-generating customer seating. Capital requirements decrease because inventory investment disappears.
Customer satisfaction increases through reduced wait times, personalized experiences, and consistent quality. Higher satisfaction drives repeat business, positive reviews, and word-of-mouth marketing. Customer lifetime value increases while acquisition costs decrease.

The cumulative impact: Restaurants with AI optimization achieve 40-50% higher profit margins than traditional operations while providing superior customer experiences. These aren't incremental improvements—they're competitive advantages that traditional operations cannot match.

But the transformation extends beyond individual restaurants to entire business ecosystems. Suppliers optimize production based on predictive demand rather than historical orders—delivery services route vehicles based on AI optimization rather than manual scheduling. Payment processors handle seamless transactions rather than terminal-based processing.

Your consulting expertise becomes valuable because you understand how these transformations create opportunities across the entire ecosystem. Restaurants need implementation strategies, suppliers need integration capabilities, technology companies need business validation, and investors need market analysis.

The Dealer Platform Transformation

My experience managing dealer networks provides crucial insights into how AI transforms traditional distribution models into platform-based ecosystems. This transformation pattern applies across industries, making the restaurant case study universally relevant.

Traditional dealer relationships operate on product distribution: manufacturers create products, dealers purchase inventory, dealers sell to customers, and dealers provide support and service. This model creates information silos, inventory risks, and limited scalability.

AI enables platform transformation where dealers become value-creation partners rather than product distributors. Instead of purchasing inventory, dealers share revenue based on customer success metrics. Instead of providing reactive support, dealers offer proactive optimization services powered by AI analytics.

Here's how the transformation works: AI systems monitor restaurant performance across all dealer locations, identifying optimization opportunities and best practices. Successful implementations automatically share knowledge across the network. Dealers receive real-time alerts about customer issues before problems impact operations. Revenue sharing aligns dealer success with customer success rather than product sales.

The economic benefits compound for all participants. Dealers reduce inventory investment and storage costs while increasing revenue through ongoing optimization services. Restaurants receive proactive support and continuous improvement rather than reactive problem-solving. Manufacturers gain real-time market feedback and reduced support costs through automated systems.

Your platform vision for hospitality creates a model for other industries. Automotive dealers become ongoing service providers rather than vehicle sellers. Technology distributors become implementation partners rather than product resellers. Consulting networks become collaborative knowledge platforms rather than individual expertise providers.

Competitive advantage comes from understanding that platform economics creates winner-take-all dynamics. The first organizations to successfully implement platform transformations gain network effects that become increasingly difficult for competitors to overcome.

Economic Impact on Business Models

The financial transformation of restaurant operations through AI provides measurable proof that these changes represent fundamental economic shifts rather than incremental improvements. The numbers eliminate doubt about the inevitability of transformation.

Automation and predictive maintenance decrease support costs by 60-70%. Traditional POS systems require extensive technical support because they're complex and failure-prone. AI-powered systems become self-optimizing and self-healing, requiring minimal human intervention once deployed at scale.

Through personalized experiences and reduced friction, customer satisfaction increases by 40-50%. Higher satisfaction translates directly to increased revenue through repeat business and positive word-of-mouth marketing. Customer lifetime value increases while acquisition costs decrease.

Dealer effectiveness improves by 50-60% through AI-powered analytics and optimization tools. Dealers shift from reactive problem-solving to proactive value creation. Revenue-sharing models based on customer success align everyone's incentives toward continuous improvement.

Scalability increases exponentially through platform economics rather than linear growth models. Traditional restaurant expansion requires duplicating systems, training, and management at each location. Platform-based operations automatically share knowledge, optimization, and resources across all locations.

The cumulative economic impact creates sustainable competitive advantages that traditional operations cannot match. AI-optimized restaurants operate with lower costs, higher quality, and better customer experiences. These advantages compound over time as AI systems continue learning and improving.

Universal Transformation Principles

The restaurant revolution reveals transformation patterns that apply across all industries. Understanding these principles enables strategic positioning regardless of your specific sector.

Conservative industries follow predictable adoption patterns: resistance based on tradition, economic pressure forcing consideration, pilot testing to prove concepts, rapid adoption once benefits become a clear, competitive advantage for early adopters, and obsolescence for late adopters.

Economic pressure drives technology acceptance faster than convenience or efficiency arguments. Organizations change when survival requires adaptation, not when technology becomes available. The most successful implementation strategies focus on measurable ROI rather than technological capabilities.

Integration challenges are technical, not conceptual. Most organizations understand that AI will transform their operations. The difficulty lies in practical implementation: legacy system integration, staff training, process redesign, and change management. Success requires understanding both technology and business operations.

Platform thinking applies equally to physical and digital industries. Manufacturing, retail, logistics, healthcare, education, and professional services all benefit from platform economics. The key insight is recognizing that platforms create value through network effects rather than individual efficiency improvements.

This analysis clarifies your cross-industry consulting applications. By understanding the underlying patterns rather than industry-specific details, you can apply hospitality transformation lessons to any sector. Resistance patterns, adoption strategies, implementation roadmaps, and economic cases translate across industries.

The competitive advantage comes from pattern recognition across domains. While others see industry-specific problems, you see universal transformation principles. While others focus on technology capabilities, you focus on business outcomes. While others sell products, you create platforms.

Key Takeaways

The restaurant revolution demonstrates universal principles that drive AI transformation across all industries and create the foundation for the Thought Economy:

- **Pattern Recognition:** Conservative industries follow predictable AI adoption curves - resistance, economic pressure, pilot testing, rapid adoption, competitive advantage. Understanding this pattern enables strategic positioning regardless of your industry
- **Platform Thinking:** The shift from product distribution to value-creation partnerships applies across all industries. Success comes from creating network effects rather than individual efficiency improvements
- **Economic Transformation:** AI optimization creates 40-50% higher profit margins while improving customer experience. These aren't incremental improvements—they're competitive advantages that traditional operations cannot match
- **Thought Economy Preview:** When operations automate, human insight and relationship-building become the primary competitive differentiators. Your consciousness becomes the valuable commodity that AI cannot replicate

Strategic Implication: The restaurant case study reveals how industry-specific expertise translates into platform economy participation. Your deep understanding of transformation patterns, implementation challenges, and success factors becomes valuable currency for communities navigating similar changes across multiple sectors.

The Bridge Forward: Restaurant automation demonstrates the economic forces creating spatial inversion—when efficiency no longer requires physical proximity, human settlement patterns begin their most dramatic reorganization in 225 years.

Chapter 8: The Reverse Industrial Revolution

When AI Reverses 225 Years of Human Civilization

For 225 years, every significant technological advancement has driven urbanization. AI is the first technology driving people away from cities. We're experiencing the mathematical reversal of human settlement patterns, compressed from centuries to decades.

Restaurant technology taught me to recognize transformation patterns twenty years before they happen. In 2003, I predicted robotic restaurants. In 2025, they're arriving exactly as envisioned. But the robotics revolution reveals something even more profound about the future of all physical work—and the complete spatial reorganization of human civilization.

The Prophetic Arc: From Natural Disasters to Mathematical Certainty

Some patterns become visible only through decades of observation, connecting disparate signals that others dismiss as isolated events. This chapter represents the mathematical validation of intuitions that formed thirty-five years ago—when I first witnessed what we now recognize as preview events of humanity's most significant spatial reorganization since agriculture.

The story begins with infrastructure failure as a prophecy. In 1989, the Loma Prieta earthquake struck the San Francisco Bay Area, collapsing freeways and disrupting transit systems. I moved to the USA about a year later during the recovery period. I witnessed something unprecedented: a major city experiencing massive workforce dispersion driven by practical necessity rather than choice.

For six months plus, San Francisco became an accidental laboratory for remote work. Downtown parking revenue collapsed, business district foot traffic plummeted, and municipal revenue streams dried up almost overnight.

The city's response was telling: they introduced a "working from home tax," desperately attempting to capture revenue from workers who were no longer contributing to the urban economic ecosystem through their physical presence.

However, my understanding of urban economic fragility deepened through an even more dramatic preview event. In 1998, while living in Auckland, New Zealand, I witnessed the central business district lose power for five weeks when four underground power cables failed due to poor maintenance. The statistics were devastating: 70,000 workers were affected, only 20% of retailers remained open, and retail trade slumped by 90%.

The crucial insight came not during the blackout but after the power returned: many offices stayed empty permanently. Businesses forced to relocate during the crisis discovered they could operate effectively from distributed locations and never returned to the CBD. Over time, many of those abandoned office buildings were converted to apartments as the economics of urban commercial space fundamentally shifted.

The pattern became unmistakable: infrastructure disruption forces business dispersion, which reveals that physical proximity was unnecessary, leading to permanent spatial reorganization. In 1989, it was earthquake damage. In 1998, it was a power failure. In 2025, it's AI making physical proximity economically irrational.

The Vision That Was Twenty Years Too Early

This early experience with multiple spatial workforce disruptions informed my later pattern recognition. When I pitched robotic restaurant franchises to VCs in 2003, I had already witnessed how quickly urban economic models could collapse when workers dispersed. The technology concept seemed premature, but the underlying understanding of urban fragility was grounded in lived experience.

I walked into venture capital offices in Sydney with detailed plans for fully automated robotic restaurant franchises. The technology concepts were sound, and the automation was feasible. Still, the outcome was a consistent response that would echo for the next two decades: "Interesting technology, but where's the problem you're solving?"

The VCs were right—for 2003. Cities were booming with knowledge workers, office districts bustled with lunch crowds seeking quick service, and human labor seemed infinite and affordable. Labor costs were manageable, customer expectations centered on human interaction, and foot traffic was strong enough to support traditional staffing models. My robotic restaurant concept was the classic "solution desperately seeking a problem."

My 2003 vision encompassed complete operational redesign around robotic capabilities rather than incremental automation of existing processes. Kitchen operations would be designed from the ground up for robotic precision and consistency. Order management would flow through systems that optimize preparation sequences. Customer service would combine automated efficiency with personalized experiences delivered through technology rather than human interaction.

One of the most innovative aspects was the container deployment strategy—using standardized shipping containers as complete restaurant units that could be deployed globally with minimal local customization. Complete restaurant operations designed within 20-foot or 40-foot shipping containers would enable factory-level manufacturing quality control. Every system—robotic food preparation, AI order management, customer interface technology, and kitchen equipment—would be precisely configured and tested before deployment.

The technical challenges of 2003 were formidable. Computing power for real-time robotics control was expensive and limited. Sensors and actuators required for food preparation precision cost more than most restaurants generate in revenue. No AI training environments existed for teaching robots complex food service tasks. Manufacturing costs for custom robotic systems were prohibitive for anything but the most significant corporate deployments.

Yet I continued developing the concepts and frameworks, understanding that technology curves would eventually enable the vision. I was building expertise in preparation for technological convergence—creating solutions for problems that didn't yet exist but would inevitably emerge.

I didn't anticipate how the spatial inversion would create the perfect economic storm, making robotic restaurants viable and essential for urban survival.

The Pattern That Changed Everything

Twenty years later, in May 2023, I published "Code Red: The Inescapable Clarion Call for a Technological Tsunami." Something had shifted in the technological landscape that made me issue an urgent warning: "Like a ticking time bomb, the AI revolution is imminent, threatening to either propel businesses into a new era of efficiency and growth or topple those who fail to adapt to its force."

What I sensed intuitively in 2023, we can now measure with logarithmic precision in 2025. The "tsunami" wasn't just about job displacement—it was about the complete spatial reorganization of human civilization.

For 225 years, every significant technological advancement has followed the same spatial pattern, driving urbanization from 3% in 1800 to 58% today. The steam engine pulled people from farms to factory towns. Electricity concentrated them in industrial cities. The automobile spread them to suburbs while keeping them tethered to urban centers. Computers and the internet initially promised dispersion but ultimately created mega-cities of knowledge workers.

Each wave of innovation increased urbanization because value creation required proximity. Steam engines needed workers near factories. Electric power plants served concentrated populations most efficiently. Automobiles enabled suburban expansion but still required commuting to central business districts. Despite their apparent location independence, even computers created Silicon Valley and similar tech clusters where innovation happened through face-to-face collaboration.

Urbanization wasn't a coincidence—it was an economic necessity until now.

The Mathematical Precision of Reversal

Working with data from the World Economic Forum's 2025 Future of Jobs Report, I developed a logarithmic displacement model that reveals something extraordinary. The same mathematical patterns that drove 225 years of urbanization are now operating in reverse, compressed from centuries to decades.

The model shows:

- **Historical Pattern:** 80-90% agricultural displacement over 100+ years during industrialization
- **Current Pattern:** 8-15% office-based displacement over 10-15 years during AI transformation
- **Validation:** 98.2% accuracy against WEF's prediction of 92 million displaced jobs from the 8.5 billion global population

The mathematics are identical, but the timeline has compressed by a factor of 10-15x. Why? Unlike physical infrastructure that requires decades to build, cognitive work scales instantly. A factory took decades to develop and automate, while an AI model deploys globally in hours, not years.

This opportunity to scale represents the first reversal of a human settlement pattern over two centuries. We're not just witnessing technological change—we're documenting the spatial reorganization of human civilization.

Tesla's Manufacturing Revolution: The Validation Model

Twenty years after my initial robotic restaurant concept, Tesla's manufacturing revolution provided the validation I needed. Elon Musk's approach to automotive production demonstrated exactly the principles I had envisioned for food service: complete operational redesign around robotic capabilities rather than automation of existing human processes.

Tesla didn't just automate traditional car assembly—they reimagined manufacturing from the ground up. Software-defined production lines adapt in real time to different vehicle configurations. Robotic precision and consistency achieve quality levels impossible with human assembly. AI-driven optimization continuously improves processes, learning from every vehicle produced and sharing improvements across all facilities globally.

The parallel to my restaurant vision was unmistakable. Tesla proved that complete process redesign, not incremental automation, creates sustainable competitive advantages. Their approach demonstrated how robotic systems achieve superior outcomes when operations are designed around mechanical capabilities rather than constrained by human limitations.

The Tesla manufacturing model validated several critical insights from my 2003 research:

Complete Workflow Redesign: Tesla redesigned every aspect of vehicle assembly around robotic capabilities. My restaurant concept similarly redesigned food preparation, order management, and customer service around automated systems rather than trying to replicate human actions.

Platform Thinking: Tesla's standardized processes enable rapid scaling across multiple facilities. The restaurant concept emphasized standardized workflows designed to be replicated globally through containerized deployment models.

Continuous Learning: Tesla's AI systems learn from every vehicle produced, constantly optimizing processes. Restaurant robotics would similarly learn from every order, continuously improving preparation speed, quality, and customer satisfaction.

Economic Transformation: Tesla's approach reduces costs and enables capabilities that are impossible to achieve with traditional manufacturing. Robotic restaurants would similarly provide consistency, speed, and quality impossible with human-only operations.

The Technical Convergence: From Impossible to Inevitable

The transformation of my 2003 vision from technically impossible to economically inevitable resulted from convergence across multiple technology domains, perfectly timed with the spatial inversion creating economic necessity.

AI Training Environment Revolution: In 2023, NVIDIA's Isaac Sim platform enabled virtual robot training, solving the biggest challenge in robotics deployment. What previously required years of physical training—teaching robots to navigate complex food preparation tasks—now happens in days through virtual simulation. Robots can practice millions of scenarios in virtual environments before ever touching real equipment, dramatically reducing deployment time and cost.

Robotic Hardware Maturation: The component costs that made my 2003 concept economically prohibitive have dropped to commercial viability. Sensors providing the precision required for food preparation now cost hundreds rather than thousands. Actuators capable of delicate manipulation are ideally available. Computing power for real-time control is available in compact, affordable packages.

AI Capability Convergence: The artificial intelligence required for dynamic restaurant operations—managing variable ingredients, adapting to customer preferences, and optimizing preparation sequences—now matches the requirements I outlined two decades ago. Computer vision systems can assess food quality and presentation with a precision that exceeds human capabilities. Natural language processing enables sophisticated customer interactions. Machine learning optimizes operations in real-time based on performance data.

Manufacturing Process Validation: Tesla's success proved that complete operational redesign around robotic capabilities creates superior outcomes rather than just cost savings. This approach validated my assumption that robotic restaurants should undergo entirely new operational models rather than automated versions of human-staffed kitchens.

The Cascade Already Begun

The evidence of spatial inversion surrounds us, hiding in plain sight:

Commercial Real Estate Collapse:

- San Francisco office vacancy: 35.8% (Q1 2025)—higher than the dot-com crash
- Manhattan office availability: up 74% since March 2020
- 74% of Fortune 500 CEOs planning office space reduction
- $1.2-2.7 trillion in commercial real estate value at risk

Population Movement Patterns:

- 58% of households leaving city centers moved further within metropolitan areas
- Small cities and rural areas experienced 1-2% population growth (2020-2022)
- Mixed-use suburban developments increased by 20% in 2023

- Remote work reduces transport emissions by 40%+

Corporate Workforce Planning:

- 41% of companies worldwide plan AI-driven workforce reductions by 2030
- 70% of organizations implementing structured automation in 2025
- 77% planning to reskill/upskill workers for distributed work models

The World Economic Forum's 2025 Future of Jobs Report reveals something profound: 92 million jobs were displaced, but 170 million were created. That's not just transformation—it's geographic liberation. Those 170 million new jobs don't need to be in the same places as the 92 million displaced ones.

The Economic Necessity: When Survival Drives Innovation

While technology convergence made robotic restaurants technically feasible, the spatial inversion created the economic conditions that made them essential for survival. The mathematics are brutal and compelling.

The ROI Reality Check:

The average restaurant worker's salary ranges from $35,000-45,000 USD annually, but that's just the beginning of labor costs. Benefits, training, scheduling complexity, sick days, turnover replacement, and management overhead easily double the price of human staffing. A comprehensive robotic kitchen system costing $100,000-150,000 USD provides 3—to 4-year payback periods under normal conditions.

However, the spatial inversion has eliminated "normal conditions" permanently.

Urban Economic Collapse:

San Francisco's 35.8% office vacancy represents 35.8% fewer lunch customers for downtown restaurants. Manhattan's 74% increase in office availability means similar customer base erosion. Restaurants that built business models around feeding knowledge workers face economic extinction as their primary customer base disperses to work remotely.

Operational Advantages Under Pressure:

The operational advantages of robotic systems become essential rather than optional when customer volumes decline dramatically:

- **Consistency:** Perfect execution every time eliminates the quality variability that drives customer dissatisfaction when every remaining customer matters for survival.
- **Speed:** Optimised movements and parallel processing tasks reduce wait times, crucial for reduced customer volume. This means every workflow is optimized for a smaller operation space.
- **Quality:** Precision measurements and timing ensure every order meets exact specifications, eliminating waste when margins become critically important.
- **24/7 Operation:** Continuous operation capability maximizes revenue opportunities from reduced customer volumes

The Container Deployment Strategy: Global Scalability Through Standardisation

The container deployment strategy, which seemed ahead in 2003, now provides the perfect model for rapid scaling in the post-spatial-inversion economy.

Standardized Manufacturing:

Complete restaurant operations designed within 20-foot or 40-foot shipping containers enable factory-level manufacturing quality control. Every system—robotic food preparation, AI order management, customer interface technology, and kitchen equipment—is precisely configured and tested before deployment, eliminating the variability and complexity of traditional restaurant buildouts.

Rapid Deployment:

Container restaurants can be deployed anywhere with basic utility connections—electricity, water, waste management, and internet connectivity. Site preparation requirements are minimal compared to traditional restaurant construction. A completely automated restaurant can be operational within days rather than months, crucial for capitalizing on market opportunities in rapidly changing urban environments.

Global Franchise Potential:

Standardised container deployment enables true global franchise scalability. Every location provides identical operational capabilities, customer experiences, and economic performance. Franchise partners receive proven systems rather than concepts requiring local adaptation. This approach dramatically reduces franchise risk while accelerating global expansion.

Economic Efficiency:

Container deployment eliminates most traditional restaurant overhead—no rent for complex facilities, minimal buildout costs, reduced permitting complexity, and lower insurance requirements. The entire economic model shifts from high-overhead operations requiring large customer volumes to efficient systems that remain profitable with reduced traffic.

Flexibility for Spatial Inversion:

As spatial inversion continues dispersing populations from urban centers to suburban and rural areas, container restaurants provide the flexibility to follow population movements. Traditional restaurants require significant investment in fixed locations. Container units can relocate to serve emerging population centers as demographic patterns shift.

The Infrastructure Enabler: Starlink's Accidental Role

The technical feasibility of the spatial inversion depends entirely on robust, low-latency internet connectivity reaching areas previously underserved by traditional broadband. The pending requirement creates an unexpected alignment between Elon Musk's Starlink project and AI-driven workforce dispersion.

Starlink's current capabilities provide exactly the infrastructure requirements for AI-enhanced distributed work:

- Download speeds of 100-200 Mbps (sufficient for cloud-based AI tools and video collaboration)
- Latency of 20-40ms (approaching fiber-optic performance for real-time AI workflows)
- Global coverage independent of terrestrial infrastructure
- Self-installation capability without requiring technician visits

The ironic alignment is profound: while Musk builds urban-concentrated manufacturing facilities like Tesla's Gigafactories, his Starlink network simultaneously enables the distributed work models that make urban concentration less necessary. Rural areas gain access to urban-level connectivity while AI provides urban-level capabilities, effectively eliminating traditional geographic advantages.

The Three Phases of Spatial Reorganisation

Phase 1: The Office Exodus (2025-2027)

The current phase involves the rapid abandonment of traditional office spaces as AI eliminates the need for physical proximity in knowledge work. Companies discover that AI-enhanced remote workers often outperform in-office teams, accelerating the transition beyond current business models as temporary remote work arrangements grow.

Key indicators:

- Office vacancy rates exceeding 30% in major metropolitan areas
- Corporate real estate portfolios shrinking by 15-35%
- Transportation usage declining in business districts
- Rural broadband infrastructure investment accelerating

Phase 2: The Great Dispersion (2027-2030)

The mathematical model predicts an 8-15% decline in urban population as knowledge workers relocate to areas with lower living costs and higher quality of life. This shift represents the first sustained decline of the urban population since the Industrial Revolution began.

Projected impacts:

- $2.7 trillion commercial real estate value destruction
- 40% reduction in commuter transportation demand
- Massive environmental benefits from reduced urban energy consumption
- Rural economic revitalization after decades of decline

Phase 3: The New Equilibrium (2030-2035)

A stable distributed economy emerges where geographic location becomes irrelevant mainly for value creation. Cities transform from work centers to experience and culture centers, while economic productivity spreads across previously underutilized geographic areas.

Characteristics:

- AI-native distributed work as the default model
- Cities specializing in entertainment, culture, and high-touch services.
- Rural areas achieve economic parity through digital connectivity.
- Transportation networks optimized for leisure rather than commuting.

The Second Wave Automation: Cities as Testing Grounds

The spatial inversion creates a two-wave automation pattern that transforms cities into laboratories for advanced automated operations while rural and suburban areas maintain human-scale economies.

Wave 1: Office Displacement (2025-2027):

Knowledge workers transition to distributed remote work, creating structural vacancy in commercial office spaces. Urban foot traffic decreases by 30-40% as daily commuters disappear. Service businesses face immediate revenue pressure as their primary customer base disperses.

Wave 2: Survival Automation (2026-2029):

Remaining urban businesses must completely automate or close permanently. Hospitality, retail, and service sectors face automation-or-extinction decisions. Traditional staffing models become economically unsustainable with reduced customer volumes. Cities evolve into testing grounds for fully automated operations.

The economics driving Wave 2 automation are inescapable. A restaurant that loses 35% of its customer base to remote work cannot maintain traditional staffing models. The choice becomes stark: implement complete automation—robotic kitchens, automated ordering, AI-managed operations—or close permanently.

Darwinian Business Evolution:

This creates Darwinian selection in real time. The businesses that survive in post-spatial-inversion cities will represent Earth's most advanced automated operations. They serve as proof-of-concept laboratories for technologies that eventually spread globally. Every operational challenge—from complex food preparation to customer service—gets solved through automation necessity rather than technological curiosity.

The Environmental Dividend

Spatial inversion offers the most significant potential environmental benefit of any technological transition in human history. The logarithmic model suggests a 40% reduction in commuter transportation could eliminate 15-25% of urban carbon emissions.

Consider the compounding effects:

- Reduced office energy consumption (25-60% decrease in commercial building demand).
- Massive reduction in transportation emissions.
- Decreased urban heat island effects.
- Lower infrastructure maintenance requirements.
- Reduced resource concentration and waste generation.

Unlike other environmental solutions that require sacrifice or behavior change, spatial inversion delivers ecological benefits as a side effect of economic optimization.

The Real Estate Reckoning

The commercial real estate implications extend far beyond office buildings. When 35% of office space becomes unnecessary, the cascading effects reshape entire urban economies:

Direct Impact:

$1.2-2.7 trillion in commercial property value destruction.

Secondary Impact: Reduced demand for urban retail, restaurants, and services.

Tertiary Impact:

Municipal revenue collapses from decreased property taxes.

Quaternary Impact:

Infrastructure overcapacity and maintenance cost burdens

Some cities will adapt by transforming commercial districts into mixed-use residential and cultural spaces, following the Auckland 1998 precedent of empty office buildings becoming apartments. Others will face economic contraction as their fundamental purpose—concentrating knowledge workers—disappears.

The Governance Challenge: 1989 Foreshadows 2025

The spatial inversion creates unprecedented challenges for governance systems designed around geographic concentration. The 1989 San Francisco earthquake previewed the jurisdictional conflicts now emerging globally.

When 15% of a city's population disperses to rural areas, how do you maintain municipal services, fund infrastructure, and preserve social cohesion? San Francisco discovered in 1989 that cities don't adapt gracefully to workforce dispersion—they fight to maintain revenue streams through taxation and regulation.

Current jurisdictional battles emerging include:

- Remote work taxes (already implemented in several jurisdictions)
- Economic presence taxes for distributed businesses
- Municipal attempts to tax companies based on remote employee locations
- Interstate conflicts over tax jurisdiction for AI-enhanced workers
- International Disputes Over Digital Nomad Taxation

The 1989 "working from home tax" was a canary in the coal mine. What San Francisco pioneered out of earthquake-driven desperation, cities worldwide are now implementing as permanent policy responses to AI-driven workforce dispersion.

The Investment Evolution: From Premature to Perfectly Timed

The investment landscape has fundamentally shifted since my 2003 experience with robotic restaurant franchises. What VCs considered "interesting but unnecessary" technology has become an essential infrastructure for business survival in the post-spatial-inversion economy.

Current Investment Flows:

Rural/Suburban Growth Capital:

- Rural broadband infrastructure projects
- Suburban mixed-use developments (+20% in 2023)
- Urban-to-rural migration services and technologies
- Distributed work collaboration platforms
- Climate-controlled rural co-working facilities

Urban Automation Acceleration:

- Robotic quick service franchises (my 2003 concept, finally economically viable)
- Fully automated hospitality management systems
- Autonomous retail operations and inventory management
- AI-managed building systems and security
- Cashless, staffless business operational models

The Auckland Conversion Opportunity:

- Commercial-to-residential conversions following the 1998 precedent
- Large, cheap office spaces transformed into upmarket apartments
- Urban districts evolving from business centers to residential communities

The Irreversible Threshold

We may have already reached the point of no return. When 35% of office space is vacant, and 41% of companies plan workforce reductions, market forces become self-reinforcing. Empty offices reduce urban vitality, making remaining businesses question their location decisions. As more companies

embrace distributed models, the competitive pressure on traditional office-based organizations intensifies.
The spatial inversion isn't a future possibility—it's a current reality accelerating toward an inevitable conclusion.

The Full Circle: Vision Validated by Necessity

My 2003 robotic restaurant concept, dismissed as "solution seeking a problem," has become the solution to problems created by spatial inversion. The economic conditions that make urban automation essential rather than optional have validated every aspect of the original vision: complete operational redesign, container deployment strategy, global standardization, and technology-first customer experience.

The transformation from "too early" to "essential for survival" demonstrates how technological vision must align with economic necessity. The spatial inversion created the monetary conditions that make robotic restaurants viable and superior to traditional operations across every meaningful metric.

Twenty years of persistence, research, and development have positioned this concept perfectly for the moment when silicon intelligence drives steel automation out of economic necessity rather than technological curiosity. The businesses that deploy these systems during the survival automation phase will become the most advanced automated operations on Earth—and the foundation for the next phase of human economic evolution.

Preparing for the Unthinkable

The spatial inversion represents the most significant reorganization of human settlement patterns since agriculture enabled permanent settlements 10,000 years ago. Unlike previous transitions that unfolded over centuries, this transformation compresses into 10-15 years.

Organizations that recognize and prepare for distributed value creation will capture enormous advantages. Those that cling to geography-dependent models risk obsolescence as competition emerges from unexpected locations with lower costs and AI-enhanced capabilities.

For individuals, spatial inversion offers unprecedented freedom to optimize life choices around preference rather than proximity to employment opportunities. The question becomes: Where would you choose to live if location didn't matter for your career?

The New Human Geography

Two hundred twenty-five years after the Industrial Revolution began concentrating humans in cities, the AI Revolution is dispersing them across the landscape. The same mathematical precision that drove urbanization now drives its reversal, compressed from centuries to decades by the exponential nature of AI deployment.

The prophetic arc completes itself across multiple decades of observation: witnessing the 1989 earthquake's preview of workforce dispersion and municipal revenue collapse; observing the 1998 Auckland power crisis and permanent commercial-to-residential conversion; developing a 2003 business concept that was "too early" but addressed real spatial economic challenges; issuing a 2023 Code Red warning that sensed approaching transformation; and arriving at 2025 mathematical models that prove the spatial inversion has begun.

What venture capitalists dismissed as "solution seeking a problem" has become essential infrastructure for business survival in emptying cities. What San Francisco experienced temporarily in 1989, we're experiencing permanently worldwide. The patterns were always there for those willing to connect disparate signals across decades of technological and economic evolution.

We stand at the inflection point of human spatial organization. The patterns that shaped civilization for two centuries are inverting before our eyes, creating opportunities and challenges unprecedented in human history.

However, success in automation creates an unexpected challenge: how do you explain workforce reduction to stakeholders without triggering resistance? When McDonald's announced AI would 'reduce employee stress,' they inadvertently revealed every company's secret playbook for selling automation as an enhancement...

Key Takeaways:

Mathematical Precision: Urbanization patterns that took 225 years to develop are reversing in 10-15 years due to AI acceleration

- **Three-Phase Evolution:** Office Exodus (2025-2027), Great Dispersion (2027-2030), New Equilibrium (2030-2035)
- **Infrastructure Enabler:** Starlink provides the connectivity that makes distributed work viable globally
- **Thought Economy Opportunity:** Geographic liberation enables global participation in consciousness trading platforms

Chapter Bridge to The Corporate Deception:

However, success in automation creates an unexpected challenge: how do you explain workforce reduction to stakeholders without triggering resistance?

When McDonald's announced AI would 'reduce employee stress,' they inadvertently revealed every company's secret playbook for selling automation as workforce enhancement rather than workforce elimination.

The spatial inversion creates the economic necessity for automation, but corporate communications have developed sophisticated strategies to mask this reality. Understanding these linguistic patterns becomes essential for navigating the transformation honestly.

Chapter 9: The Corporate Deception

Why They're Not Reducing Stress

When McDonald's announced AI would 'reduce employee stress' by eliminating decisions, they accidentally revealed every company's secret playbook for selling workforce elimination as workforce empowerment.

When McDonald's Told Us the Truth by Accident

McDonald's recent announcement about deploying AI across its 43,000 locations caught my attention—not for what they're doing, but for how they're selling it. Their messaging promised that AI would "reduce employee stress" by eliminating difficult decisions like "Did I prepare enough chicken?" or "What order goes to which car?"

With over thirty years of implementing technology solutions and building POS systems for hospitality, I recognized this language immediately. Having designed systems automating decision-making processes, I understood what McDonald's described: the systematic elimination of human judgment and decision-making from restaurant operations.

It's not about helping employees—it's about needing fewer of them.

That revelation sent me down a research rabbit hole that exposed something more sophisticated and disturbing: a coordinated corporate playbook for selling workforce elimination as workforce empowerment. Every industry, every company, and every AI deployment follows the same linguistic patterns, using carefully crafted euphemisms to mask the reality of job displacement.

The spatial inversion we explored in Chapter 7 creates the economic conditions that make automation necessary for urban business survival. The robotic systems detailed in Chapter 8 provide the technical capabilities to replace human workers entirely. However, a communication challenge lies between economic necessity and technological capability: How do organizations explain workforce reduction to stakeholders, employees, and the public without triggering resistance or backlash?

The answer is the corporate euphemism playbook—a systematic approach to linguistic misdirection that transforms "reducing employees" into "reducing employee stress," "workforce replacement" into "workforce optimization," and "job elimination" into "freeing workers for higher-value tasks."

McDonald's accidentally revealed this game by being too explicit about what they meant by "reducing stress." They told us the truth when they meant to tell us the story.

The Technology Architect's Perspective: Uncomfortable Knowledge

This analysis becomes intensely personal because of my unique position at the intersection of technological capability and business implementation. As a solutions consultant working across multiple business models—from corporate enterprise to hospitality systems—I've witnessed the evolution of technology deployment from design to implementation for over thirty years. My career has spanned hardware and software development, giving me a comprehensive understanding of technology and how AI systems operate, built, and deployed across industries. I understand the technical elegance of automated decision-making solutions and the genuine business pressures driving their adoption. Labor costs are rising, and AI capabilities are expanding. The return on investment (ROI) from efficiency gains is measurable and is now considered a competitive advantage.

But having designed and deployed systems that eliminate human decision-making, I see both the Promise and the peril of our linguistic games. I know what "reducing employee stress" means in system design terms: removing human variability, eliminating judgment calls, automating responses, and creating predictable outcomes that don't require human intervention.

When McDonald's talks about eliminating the decision "Did I prepare enough chicken?", they describe the automation of inventory management, demand forecasting, and production planning. These are sophisticated cognitive tasks that, when automated, eliminate the need for experienced human judgment in restaurant operations.

I understand why companies use euphemisms to communicate these changes, but I know what they're implementing. The gap between public messaging and technical reality has become so vast that it constitutes systematic deception.

The Corporate Euphemism Playbook

After analyzing dozens of corporate communications, earnings calls, and AI deployment announcements across multiple industries, I've identified the systematic language strategies companies use to mask workforce displacement:

The "Digital Workforce" Deception

Companies like Salesforce now describe providing a "digital workforce" where humans and automated agents "work together" in collaborative harmony. Platforms like Relevance AI sell an "AI Workforce" as "a digital team you can hire to assist you in completing mundane and repetitive tasks."

The language deliberately parallels human hiring terminology—recruiting, onboarding, training, managing—except these "digital employees" never ask for raises, benefits, sick days, or career advancement. They don't form unions, file complaints, or request time off. They work 24/7 without breaks, never disagree with management decisions, and can be "terminated" without severance packages or legal complications.

The deception lies in describing this as collaborative rather than replacement. The "digital workforce" doesn't work alongside humans—it performs human tasks without human involvement. The collaboration is between human managers and automated systems, not between human workers and digital assistants.

The "Human-in-the-Loop" Fiction

This euphemism refers to "human oversight, intervention, and collaboration integrated into Automated Decision Making Systems." Corporate communications emphasize human involvement and control, suggesting meaningful collaboration between humans and AI.

The reality often reduces humans to monitoring what AI systems do rather than participating in meaningful decision-making. It's the difference between being a pilot actively controlling an aircraft and being a passenger watching the autopilot manage the flight.

In most "human-in-the-loop" implementations, humans serve as exception handlers—intervening only when automated systems encounter situations outside their programmed parameters. The human role becomes reactive rather than proactive, supervisory rather than collaborative.

These changes represent a fundamental shift in human work from active decision-making to passive monitoring, but the euphemism suggests enhanced human capability rather than diminished human agency.

The "Workforce Optimization" Calculation

Consulting firms like KPMG promote "workforce optimization," where "organizations must systematically orchestrate work between human and digital labor." Are we witnessing strategic resource allocation, but the optimization calculation is mathematical: determining the minimum number of humans required to achieve business objectives?

The euphemism "optimization" obscures the reality of "reduction." When organizations optimize their workforce through AI deployment, they calculate the optimal human-to-AI ratio to minimize labor costs while maintaining operational capability.

Some say it is resource allocation; others say it is cost minimization through human replacement. Mathematical optimization treats human labor as a variable to be minimized rather than a capability to be enhanced.

The "Augmentation vs. Automation" Shell Game

Perhaps the most sophisticated linguistic maneuver involves distinguishing between "augmentation" (empowering workers) and "automation" (replacing workers) in corporate messaging, even when the technical implementation and business outcomes are identical.

Companies have learned to emphasize augmentation in public communications while implementing automation in actual operations. The same AI system that automates customer service can be described as "augmenting human representatives" or "automating customer interactions," depending on the audience and communication objectives.

The shell game works because the technical distinction between augmentation and automation often disappears in practical implementation. An AI system that "augments" human customer service representatives by handling 90% of interactions effectively automates the role, regardless of the terminology used to describe it.

The Reality Behind the Messaging: What Research Shows

While corporations promote AI as empowering human workers, independent research reveals a different reality. Professor Milena Nikolova's comprehensive study of thousands of workers across 20 European countries provides crucial evidence about automation's impact on human work.

The research findings directly contradict corporate messaging about AI enhancing human capabilities:

"Freeing employees for higher-value work" vs. Reality:

Automation "increased repetitive and monotonous tasks for humans" and made "human work more routine, not less."

"Enhanced human decision-making and autonomy" vs. Reality:

Workers became "more reliant on a machine's pace of work and had a more limited understanding of the full production process."

"Improved job satisfaction through reduced drudgery" vs. Reality:

Automation created more constrained work environments with less human agency.

The gap between corporate promises and research findings reveals the systematic nature of the deception. Companies aren't just optimistic about AI's impact—they're actively misrepresenting what happens when automation is deployed in real work environments.

Current Industry Examples: The Pattern Accelerating

The pattern of euphemistic communication around workforce displacement is accelerating across industries as AI capabilities expand and economic pressures increase:

Media and Content:

In 2020, MSN replaced human journalists with AI, describing the move as "content optimization" rather than layoffs. The AI systems now generate news articles, sports summaries, and financial reports without human involvement.

Education Technology:

Duolingo cut 10% of contractors in favor of AI translation systems, framing it as "improving educational effectiveness" rather than workforce reduction. The AI handles language content that previously required human translators and educational specialists.

Human Resources:

IBM's CEO announced plans to replace HR evaluation and assessment roles with AI systems, describing it as "enhancing human insight" while explicitly stating intentions to eliminate human positions.

Telecommunications:

BT plans to cut 55,000 jobs while replacing 10,000 positions with AI systems, promoting this as "digital transformation" and "future-ready workforce optimization."

Each example follows the same pattern: workforce reduction is described as improvement, efficiency, optimization, or enhancement rather than job elimination.

The Skills Graveyard: What We're Losing Through Systematic Deskilling

The implications of corporate deception extend far beyond job numbers to systematic deskilling—the erosion of human capabilities through disuse. As companies automate decision-making processes, they claim to be "augmenting" workers lose the cognitive skills that made them valuable.

The Learning Paradox

Research reveals that AI deployment creates a paradoxical relationship between efficiency and capability development. Studies show that AI "significantly inhibits people's on-the-job learning" because it "makes workers more pessimistic about the future, leading to burnout and less motivation for learning."

Automation systems that promise to enhance human capabilities reduce the cognitive challenges that develop those capabilities. Like the humans in WALL-E who gradually lost the ability to walk through overreliance on automated transportation, we're systematically losing cognitive abilities through overreliance on automated decision-making.

The brain is "famously excellent at forgetting," and automation reduces the cognitive exposure that keeps skills sharp; when AI systems handle complex analysis, human analytical capabilities atrophy. When automated systems manage strategic planning, human strategic thinking skills diminish.

Specific Capability Losses

The systematic shift from human decision-making to AI monitoring creates predictable patterns of skill degradation:

Writing and Communication:

From original composition to AI direction and editing. Workers cannot create coherent arguments, structure complex ideas, and communicate nuanced concepts.

Research and Analysis:

From investigation and synthesis to prompt engineering and result selection. The ability to identify relevant sources, evaluate credibility, and synthesize insights degrades when AI systems handle information processing.
Problem-Solving: From analysis and solution development to solution selection from AI-generated options. The cognitive processes that break down complex problems and develop innovative solutions atrophy when AI provides the analysis.

Critical Thinking:

From reasoning and evaluation to result verification and implementation. The ability to identify assumptions, evaluate evidence, and construct logical arguments diminishes when AI systems handle reasoning tasks.

These aren't just job skills—they're cognitive capabilities that define human intellectual development. The systematic automation of decision-making processes eliminates the mental challenges that develop and maintain these capabilities.

Professional Observations: Witnessing Skill Degradation

I've witnessed this skill degradation firsthand in my consulting work across multiple industries. Clients who previously demonstrated sophisticated analytical and strategic thinking increasingly rely on AI-generated insights without developing the capability to evaluate or improve those insights.

The pattern is consistent: initial efficiency gains from AI deployment result in a gradual reduction in human capability to perform the tasks that AI has automated. Workers become dependent on AI systems not just for efficiency but also for capabilities they once possessed independently.

Looking at the big picture, this creates a dangerous cycle: reduced human capability increases dependence on AI systems, further reducing human capability development and creating greater dependence. The euphemistic description of this process as "augmentation" masks the reality of systematic deskilling.

The 10-Year Education Scenario: A Generation Prepared for Jobs That Won't Exist

The corporate euphemism playbook doesn't just mask current workforce displacement—it obscures the long-term implications for education and career development. If current trends continue, we're educating a generation about employment opportunities that may not exist by the time they graduate.

The Bifurcated Future

McKinsey research reveals that "highly automatable activities dominate many current middle-wage jobs in advanced economies," creating "wage polarization, income inequality, and the lack of income advancement." The middle-skill jobs that historically provided career advancement opportunities are disappearing through automation.

We are creating a bifurcated employment future:

High-Skill AI Management Roles:

Requiring advanced technical literacy, strategic thinking, and complex problem-solving capabilities. These positions demand a sophisticated understanding of AI systems, data analysis, and technology implementation.

Low-Skill AI-Assisted Roles:

Requiring minimal human judgment, routine task execution, and basic technology operation. These positions involve monitoring automated systems and handling exceptions that AI cannot resolve.

The Disappearing Middle:

Where most current jobs exist, roles requiring moderate skill, experience, and judgment are now under attack by increasing automation through AI deployment.

Traditional education systems prepare students for middle-skill employment, which AI systems systematically eliminate. The four-year college model assumes that knowledge accumulation and skill development lead to career advancement opportunities that may not exist in an AI-automated economy.

The Turnitin CEO's Vision

The devaluation of traditional education and experience becomes explicit in statements like those from Turnitin's CEO, who envisions "future employees hired out of high school rather than needing multiple years of experience" because AI systems handle the complex tasks that previously required education and experience to develop.

This represents the fundamental devaluation of human development, learning, and experience accumulation. If AI systems can handle the analytical, strategic, and technical tasks that justify higher education and career development, what value does human intellectual development provide?

The corporate euphemism playbook obscures this devaluation by describing it as efficiency, optimization, and accessibility rather than systematically replacing human intellectual capability with automated systems.

Educational Institution Response

Universities and educational institutions face an impossible challenge: preparing students for an economy that's systematically eliminating the cognitive tasks that higher education develops. Traditional curricula emphasize analytical thinking, research capabilities, and problem-solving skills that AI systems increasingly handle more efficiently than humans.

The gap between academic preparation and economic reality creates a generation equipped with capabilities that may be economically irrelevant by graduation. Corporate euphemisms mask this misalignment by suggesting that AI augments rather than replaces human cognitive abilities.

Strategic Response Framework

Understanding the corporate euphemism playbook creates strategic opportunities for individuals and organizations willing to communicate honestly about AI's workforce impact.

For Individual Professionals:

The corporate euphemism playbook creates opportunities for honest communicators who can decode messaging and provide authentic analysis. In the Thought Economy, transparency becomes a competitive advantage.

Competitive Positioning Strategy:

1. **Decode corporate euphemisms** for clients who need honest assessment
2. **Provide authentic analysis** of AI's actual workforce impact
3. **Develop transition strategies** based on reality rather than marketing spin
4. **Build reputation** for honest communication in an era of linguistic manipulation

Platform Economy Integration:

- **Expertise Authentication:** Honest analysis of AI impact creates credibility in platform communities
- **Community Building:** Professionals seeking authentic engagement value transparent communication

- **Thought Leadership:** Cutting through euphemisms to explain actual implications builds valuable reputation

For Organizations

Companies that honestly acknowledge automation's workforce implications can build trust with employees, customers, and stakeholders. Honest communication about transition strategies, retraining programs, and alternative career paths creates competitive advantages over organizations that rely on euphemistic messaging.

Transparent organizations can attract talent who prefer honest communication over linguistic manipulation. They can build customer loyalty among constituencies that value authenticity over marketing spin. They can also develop more effective automation strategies because they address real challenges rather than euphemistic versions.

For Individuals

Understanding euphemistic communication patterns enables better career decision-making. When companies describe AI "augmentation," individuals can evaluate whether this means enhanced capability or reduced human involvement. When organizations promote "workforce optimization," individuals can assess whether this creates opportunities or eliminates positions.

The ability to decode corporate messaging becomes a crucial skill for navigating the AI economy. Individuals who recognize euphemisms can make strategic decisions based on actual trends rather than corporate communications designed to minimize resistance to workforce changes.

For Society

Demanding honest conversation about AI's workforce impact creates the foundation for effective policy responses. The euphemistic playbook prevents society from addressing the challenges of workforce transition, education realignment, and social safety net adaptation.

Honest communication enables productive debate about automation's benefits and costs, appropriate policy responses, and strategies for managing technological transition. Euphemistic communication prevents these conversations by misrepresenting the nature and scale of the changes occurring.

Platform Economy Integration: Transparency as Platform Value

The platform economy that emerges from spatial inversion and automation creates opportunities for individuals and organizations to prioritize honest communication over euphemistic messaging.

Expertise Authentication

In platform economies where expertise becomes currency, the ability to provide honest analysis rather than euphemistic spin becomes valuable. Consultants, analysts, and advisors who decode corporate messaging and explain actual implications build credibility that translates into platform value.

Community Building

Platform communities value authentic communication over marketing messages. Organizations and individuals acknowledging automation's real implications while developing constructive responses attract community members seeking honest engagement with technological change.

Strategic Consulting

The gap between corporate euphemisms and operational reality creates consulting opportunities for experts who understand technology and communication strategies. Organizations need guidance navigating between public messaging requirements and practical implementation challenges.

The Technology Implementer's Dilemma

My position as someone who analyzes and implements these automation systems creates an ethical challenge that extends beyond individual career considerations to fundamental questions about technological honesty and social responsibility.

I understand the business logic driving euphemistic communication: workforce reduction creates resistance, automation eliminates jobs that real people depend on, and honest communication about displacement can trigger a political and social backlash that complicates implementation.

But I also understand the long-term costs of systematic deception about technological change. Euphemistic communication prevents adaptive responses, delays necessary policy adjustments, and creates false expectations that make transition challenges more difficult when reality becomes unavoidable.

The corporate euphemism playbook allows short-term implementation of automation systems while avoiding immediate resistance, but it prevents the honest conversations necessary for effectively managing long-term technological transition.

Breaking the Euphemism Cycle

The solution to corporate deception about automation isn't opposition to technological development but honest communication about its impact. Organizations, individuals, and society benefit from an accurate understanding of automation's implications rather than euphemistic misrepresentation.

Honest Corporate Communication

Companies can maintain competitive advantages through automation while communicating honestly about workforce implications. Transparent transition strategies, realistic timelines, and genuine support for displaced workers create better outcomes than euphemistic messaging followed by surprise displacement.

Individual Strategic Response

Understanding euphemistic patterns enables strategic career decisions based on actual rather than claimed technological impact. Individuals can develop capabilities that complement rather than compete with automation, build relationships that survive the technological transition, and create value in platform economies that emerge from spatial inversion.

Societal Adaptation

Honest communication about automation enables productive policy responses, educational adaptation, and social safety net development. Society can manage technological transition more effectively when it understands AI deployment's actual rather than euphemistic implications.

The Uncomfortable Truth: Calling It What It Is

McDonald's accidentally revealed the truth when they described AI as "reducing employee stress" by eliminating human decision-making. They meant reducing employees, and every company using similar euphemisms means the same thing.

The AI revolution offers genuine benefits, but we won't realize them if we can't have honest conversations about the costs. The corporate euphemism playbook prevents that honesty by systematically misrepresenting workforce replacement as workforce empowerment.

It's time to call it what it is: workforce replacement disguised as workforce enhancement, job elimination marketed as employee stress reduction, and systematic deskilling promoted as human augmentation.
As someone who implements these systems and understands their capabilities, I've seen both sides of this transformation. The technology is genuinely impressive, the business benefits are real, and the efficiency gains are substantial. However, the euphemistic communication about workforce impact prevents the honest conversations necessary for managing this transition effectively.

The spatial inversion creates economic conditions that make automation necessary for business survival. The technology provides capabilities that make human replacement technically feasible. The corporate euphemism playbook provides linguistic strategies that make workforce displacement politically acceptable.

Understanding all three components—economic necessity, technological capability, and communication strategy—becomes essential for the transformation.

Key Takeaways

The corporate euphemism playbook reveals how organizations systematically disguise workforce transformation while creating opportunities for honest communicators in the emerging Thought Economy:

- **Corporate Euphemism Playbook:** "Reducing employee stress" = reducing employees; "workforce optimization" = workforce reduction; "digital workforce" = human replacement disguised as collaboration

- **Systematic Deskilling:** AI deployment reduces cognitive challenges that develop human capabilities, creating dependence while being marketed as empowerment
- **Competitive Advantage:** Honest communication about AI impact creates differentiation in markets filled with euphemistic messaging—authenticity becomes a premium brand
- **Thought Economy Bridge:** Authentic analysis of technological change becomes valuable currency in platform communities seeking honest insight rather than corporate spin

The Strategic Opportunity: While corporations use euphemisms to mask workforce displacement, professionals who decode these patterns and provide honest analysis build credibility that translates directly into platform economy value. Your ability to cut through linguistic manipulation becomes monetizable expertise.

What's Next: Understanding corporate deception about workforce impact is crucial, but it raises a deeper question: when expertise becomes downloadable and knowledge becomes commoditized, what justifies premium pricing for human insight? The consulting industry faces its own existential transformation.

Chapter 10: The Consultant's Dilemma

When Expertise Becomes Downloadable

The global management consulting industry generates $175 billion annually by providing access to specialized knowledge. Today's frontier AI systems deliver PhD-level analysis for $20/month. When knowledge becomes a commodity, what justifies premium pricing for human Delivery?

The $175 Billion Question

The global management consulting industry generates $175 billion annually by providing access to specialized knowledge, strategic frameworks, and implementation expertise. McKinsey, BCG, Deloitte, and hundreds of smaller firms employ PhD-level analysts who command premium rates for their ability to analyze complex business challenges and develop strategic solutions.

This model worked because expertise was scarce, expensive, and required years of education and experience to develop. Organizations paid consultants to bring external knowledge that they couldn't access internally.

Then everything changed.

Today's frontier large language models have been trained on virtually all human-written knowledge and can deliver PhD-level analysis on demand. The specialized knowledge that once required hiring a McKinsey consultant, you can now mine this knowledge by prompting an AI system for a monthly subscription fee of $20.

McKinsey's strategic frameworks? Available through AI prompts. BCG's market analysis methodologies? Downloadable through sophisticated AI queries. Deloitte's implementation plans? Generated through AI collaboration.

The democratization of expertise creates an existential challenge for traditional consulting: when knowledge becomes a commodity, what justifies premium pricing for the human Delivery of that knowledge?

My Personal Consulting Revolution

This challenge became personal when I realized how AI transformed my consulting practice. Over thirty years of consulting and technology implementation, I have built expertise in ERP migrations, POS system deployment, and business technology integration. My value proposition centered on bringing specialized knowledge that clients couldn't access internally.

The emergence of sophisticated AI systems forced a fundamental reassessment of that value proposition.

Initially, I used AI as a research assistant—helping analyze client challenges, generate implementation plans, and document best practices. The efficiency gains were substantial: tasks that previously required hours of analysis gave way to being completed in minutes with AI assistance.

But efficiency wasn't the fundamental transformation. The real change was qualitative: AI didn't just help me work faster—it helped me think better. Complex strategic frameworks became more sophisticated when developed collaboratively with AI systems that could access broader knowledge bases and simultaneously identify patterns across multiple industries.

Client delivery improved dramatically. Instead of providing solutions based on my experience and knowledge, I could offer insights that synthesized best practices across thousands of similar implementations, validated through a comprehensive analysis of successful strategies, and customized for specific client contexts.

The competitive advantage was overwhelming. Traditional consultants working without AI augmentation couldn't match the depth, speed, or sophistication of AI-enhanced analysis and strategy development.

But this advantage came with an uncomfortable realization: if AI could dramatically enhance my consulting capabilities, it could eventually replace many consulting functions entirely.

The Acceleration Timeline: From Assistant to Primary Consultant

The transformation of consulting through AI follows a predictable acceleration pattern that I've witnessed firsthand:

2023: AI as Research Assistant

AI systems handle information gathering, document analysis, and preliminary research. Consultants maintain control over strategic thinking and client relationships. The value proposition remains unchanged mainly because AI improves efficiency.

2024: AI as Analysis Partner

AI systems generate strategic frameworks, perform sophisticated analysis, and develop implementation plans. Consultants become orchestrators of AI capabilities while providing human judgment and client interaction. Value proposition shifts toward AI-enhanced insights.

2025: AI as Primary Consultant with Human Oversight

AI systems handle end-to-end consulting projects with minimal human intervention. Consultants become quality controllers and relationship managers. Traditional consulting models face fundamental disruption.

2026+: Platform Economy for Specialized Human Insights

AI systems provide standardized consulting capabilities. Human consultants survive only by offering highly specialized insights, industry-specific knowledge, or relationship-dependent services that AI cannot replicate. We're transitioning from phase 2 to phase 3, and the acceleration is exponential rather than linear.

Three Consulting Futures: Transform, Specialize, or Disappear

Every consultant, analyst, and knowledge worker faces the same strategic choice: how to respond to the commoditization of expertise through AI. Based on my analysis of industry trends and personal experience with AI transformation, three distinct futures emerge:

Future 1: Transform (The Path I've Chosen)

Strategy: Embrace AI as a cognitive multiplier rather than a competitive threat.

Approach: Combine human judgment with AI capabilities to deliver enhanced value that neither humans nor AI can provide independently.

Implementation: Develop expertise in AI orchestration, maintain direct client relationships, and focus on strategic guidance that requires human understanding of business context, stakeholder dynamics, and implementation challenges.

Transformation Success Factors:

- **Rapid AI Adoption:** Early adoption of AI tools provides competitive advantages during transition periods
- **Client Education:** Help clients understand AI capabilities while demonstrating human value-add
- **Relationship Depth:** Strengthen personal connections that AI cannot replace
- **Continuous Learning:** Develop expertise in emerging AI capabilities and integration strategies

My consulting practice exemplifies this transformation. Instead of competing with AI, I orchestrate AI capabilities to deliver insights and solutions that exceed what traditional consulting can provide. Clients receive more sophisticated analysis, faster turnaround times, and solutions validated against broader knowledge bases than any individual consultant could access independently.

The competitive advantage comes from understanding how to extract maximum value from AI systems while providing human oversight, strategic judgment, and relationship management that clients still require.

Future 2: Specialize (The Boutique Survival Strategy)

Strategy: Develop ultra-specialized expertise in narrow domains where AI capabilities remain limited.

Approach: Focus on industry-specific knowledge, regulatory expertise, or complex relationship management that requires deep human understanding.

Implementation: Build deep specialization in mergers and acquisitions, regulatory compliance, crisis management, or cultural transformation, where human judgment and relationship skills remain essential.

Specialization Success Factors:

- **Domain Expertise:** Develop knowledge depth that exceeds AI training data
- **Regulatory Understanding:** Master complex compliance requirements that AI cannot navigate independently
- **Relationship Skills:** Focus on stakeholder management and change leadership
- **Niche Focus:** Serve markets too specialized or sensitive for AI-only solutions

Specialized consultants survive by serving markets where AI cannot provide adequate solutions—either because the knowledge domain is too narrow, the regulatory requirements too complex, or the relationship dynamics too sensitive for AI-only consulting.

This strategy works for consultants willing to narrow their focus dramatically and accept smaller market opportunities in exchange for reduced AI competition.

Future 3: Disappear (The Traditional Path to Obsolescence)

Strategy: Continue traditional consulting approaches without AI integration.
Approach: Compete on human expertise and traditional consulting methodologies.
Implementation: Maintain existing service delivery models while hoping AI disruption remains limited or distant.

Disappearance Inevitability:

- **Cost Disadvantage:** Traditional consulting cannot compete with AI-enhanced alternatives on price
- **Quality Disadvantage:** Human-only analysis cannot match AI-augmented insights
- **Speed Disadvantage:** Traditional research and analysis cannot compete with AI capabilities

- **Scale Disadvantage:** Individual consultants cannot access knowledge bases equivalent to AI training data

Consultants choosing this path face inevitable displacement as AI capabilities expand and client expectations adjust to AI-enhanced service delivery. The only Question is timing: whether displacement happens in months or years depends on how quickly clients adopt AI-augmented alternatives.

The Agentic AI Revolution: Every Business Becoming IT-First

The consulting industry transformation occurs within a broader context: the emergence of Agentic AI that enables businesses to internalize capabilities that previously required external consultants.

Agentic AI—artificial intelligence designed to act autonomously with purpose and agency—represents a fundamental shift in how businesses access expertise and implement solutions. For the first time, organizations can deploy AI systems that perform end-to-end consulting functions without human intervention.

The Business Transformation Model

Consider a mid-sized logistics company that recently replaced their entire fleet management consulting engagement with an Agentic AI solution that:

- Optimizes routes and schedules in real-time
- Predicts maintenance needs with 94% accuracy
- Automatically adjusts to supply chain disruptions
- Manages customer communications during delays
- Continuously improves based on performance outcomes

Their transportation director's assessment: "We're now a technology company that happens to move freight."

This transformation pattern applies across industries. Manufacturing companies use AI to optimize production schedules, predict maintenance needs, and reconfigure supply chains autonomously. Financial institutions deploy AI systems that monitor transactions, adjust risk models, and personalize customer experiences. Healthcare providers implement AI that analyzes patient data, suggests treatment protocols, and manages resource allocation.

The Consulting Capability Integration

Agentic AI enables businesses to internalize the five core consulting functions:

1. Problem Analysis

AI systems analyze business challenges with remarkable precision, identifying root causes, market dynamics, and organizational constraints that human consultants traditionally uncover through discovery processes.

2. Vision & Strategy Development

AI generates transformation roadmaps and business blueprints, creates detailed models of future states, and develops cost-benefit analyses for different strategic approaches.

3. Solution Architecture

AI designs integrated systems leveraging existing infrastructure, creates technical specifications for new capabilities, and develops implementation sequences with minimal disruption.

4. Implementation & Integration

AI generates code, workflows, and business rules, configures systems, establishes data flows, and tests and validates functionality against requirements.

5. Continuous Optimization

AI monitors performance against key metrics, identifies improvement opportunities, and adapts to changing market conditions without human intervention.

The traditional consulting value proposition disappears when businesses can access these capabilities internally through AI deployment.

New Value Creation: From Selling Time to Orchestrating Intelligence

The consulting industry's survival depends on developing new value-creation models that transcend traditional time-for-money exchanges. Based on my experience transitioning from traditional consulting to AI-enhanced service delivery, three emerging models show promise:

Intelligence Orchestration

Instead of selling human analysis, consultants become orchestrators of AI capabilities, helping clients deploy and optimize AI systems for specific business challenges. This approach requires understanding AI capabilities, integration challenges, and strategic applications rather than domain expertise alone.

Value creation comes from knowing how to extract maximum value from AI systems while providing human oversight for strategic decision-making and stakeholder management.

Platform Development

Consultants create platforms combining AI capabilities with industry-specific knowledge, enabling clients to access technological capabilities and specialized expertise through integrated systems.

This model shifts from project-based consulting to platform-based subscription services where clients access ongoing AI-enhanced capabilities rather than discrete consulting engagements.

Community Building

Consultants develop communities around AI-enhanced problem-solving, bringing together clients facing similar challenges and providing facilitated access to AI capabilities for collaborative solution development.

Value creation comes from community development, facilitation skills, and the ability to help clients learn from each other while leveraging shared AI capabilities.

Why Agility Beats Scale: The New Competitive Dynamics

The AI revolution creates competitive dynamics that favor agile individual consultants over large consulting firms. While traditional consulting advantages centered on scale, resources, and access to specialized talent, AI-enhanced consulting reverses these advantages.

Large Firm Disadvantages

Slow Decision-Making: Complex approval processes prevent rapid AI adoption and optimization. **Legacy Constraints:** Existing business models and client expectations resist AI transformation. **Resource Allocation:** Heavy overhead structures make AI-enhanced pricing models economically challenging. **Cultural Resistance:** Large organizations face internal resistance to fundamental business model changes

Individual Consultant Advantages

Rapid Implementation: Immediate AI tool adoption without organizational constraints.
Direct Client Relationships: Personal connections that survive the technological transition.
Flexible Pricing: Ability to adjust service delivery and pricing models quickly.
Continuous Learning: Faster adaptation to emerging AI capabilities and client needs.

My consulting practice exemplifies these advantages. While large firms debate AI strategy and navigate organizational resistance, I immediately deploy new AI capabilities and adjust service delivery based on client feedback and technological advancement.

The competitive advantage comes from speed and flexibility rather than scale and resources.

Platform Economy Positioning: Consulting Evolution as Platform Participation

The transformation of consulting from traditional service delivery to AI-enhanced capabilities creates opportunities for platform economy participation beyond individual client relationships.

Community Building Around AI-Enhanced Consulting

My evolution toward platform thinking involves developing communities around AI-enhanced business transformation, bringing together clients, technology providers, and other consultants for collaborative problem-solving and knowledge sharing.

Instead of competing for individual consulting engagements, I facilitate platforms where participants share AI tools, methodologies, and insights for mutual benefit.

Knowledge Sharing and Collaborative Improvement

Platform participation involves contributing insights, tools, and methodologies to communities that benefit all participants. Yields value knowledge sharing rather than knowledge hoarding.

Revenue diversification comes through platform engagement—subscription services, community facilitation, tool development, and collaborative projects—rather than traditional time-based consulting.

The Consulting Industry Future

The consulting industry's future involves consolidation around AI-enhanced service providers who can deliver superior outcomes at lower costs than traditional consulting while developing platform-mediated expertise distribution rather than firm-based delivery models.
Individual consultants who master AI orchestration and platform participation will capture market share from traditional consulting firms that resist technological transformation.

Community-driven knowledge development and sharing will replace proprietary methodologies and frameworks as competitive advantages.

Strategic Implications: The Three-Player Game

The consulting transformation occurs within a broader three-player competitive landscape that determines which organizations and individuals thrive in the AI economy:

Player 1: Traditional Incumbents

Large consulting firms that built competitive advantages around human expertise, proprietary methodologies, and relationship management. These organizations face the most significant disruption from AI commoditization of knowledge.

Player 2: AI-Native Competitors

Organizations and individuals that build competitive advantages around AI capabilities offer superior analysis and implementation at lower costs than traditional consulting.

Player 3: AI-Enhanced Entrepreneurs

Consultants and advisors who combine AI capabilities with human judgment, relationship skills, and platform participation to create hybrid value propositions that exceed both traditional consulting and pure AI solutions.

My strategic positioning targets Player 3: using AI enhancement to provide superior capabilities while maintaining human relationships and judgment that pure AI solutions cannot replicate.

Personal Strategic Evolution: From Solutions Consultant to Platform Participant

My consulting practice evolution provides a case study for knowledge workers navigating AI-driven industry transformation:

Phase 1: Traditional Expertise (1990s-2010s) Built competitive advantages around specialized knowledge in enterprise systems, ERP deployment, and POS system integration—value creation through knowledge scarcity and implementation experience.

Phase 2: AI-Enhanced Delivery (2020s)

Integrated AI capabilities to improve analysis speed, solution quality, and client outcomes. Maintained traditional consulting relationships while enhancing service delivery through AI augmentation.

Phase 3: Platform Participation (Current)

Developing communities around AI-enhanced business transformation, contributing to shared knowledge platforms, and creating value through collaborative problem-solving rather than individual consulting engagements.

Phase 4: Platform Leadership (Future)

Contributing to platform governance, methodology development, and community building while maintaining specialized expertise in spatial inversion, automation strategies, and business transformation.

This evolution demonstrates how individual consultants can navigate AI disruption by embracing rather than resisting technological change while developing new value-creation models that survive AI commoditization of traditional expertise.

The Speed Advantage: Why Individual Consultants Will Outcompete Large Firms

The AI revolution creates competitive advantages for speed and agility over scale and resources. In my consulting practice, these advantages manifest in several ways:

Immediate AI Tool Adoption:

While large firms evaluate AI strategies through committee processes, I deploy new capabilities immediately and optimize based on client results.

Direct Client Relationships:

Personal connections with decision-makers enable rapid communication and feedback cycles that large firms cannot match through account management structures.

Customized Solutions:

AI enhancement enables personalized service delivery that addresses specific client needs rather than standardized methodologies designed for broad applications.

Continuous Innovation:

Constant experimentation with new AI capabilities and integration strategies provides competitive advantages that large firms cannot match due to organizational constraints.

The result is superior client outcomes at lower costs with faster delivery times—a competitive advantage that large consulting firms cannot replicate without fundamental organizational transformation.

Preparing for the Platform Economy: Beyond Traditional Consulting

The platform economy that emerges from AI transformation creates opportunities for consultants willing to evolve beyond traditional service delivery models:

Revenue Diversification Strategies

Community Subscriptions: Monthly or annual fees for access to AI-enhanced consulting communities.
Platform Facilitation: Revenue sharing from successful project collaborations.
Tool Development: Creating and licensing AI-enhanced consulting methodologies.
Content Creation: Developing educational content around AI business transformation.

Value Creation Evolution

From Knowledge Hoarding to Knowledge Sharing: Contributing expertise to platforms for collective benefit.
From Project Delivery to Relationship Building: Developing long-term platform participation rather than discrete consulting engagements.
From Individual Expertise to Collaborative Intelligence: Combining human judgment with AI capabilities and community insights.

Competitive Positioning

AI Orchestration Skills: Expertise in maximizing value from AI systems.
Platform Development: Ability to build and facilitate collaborative communities.

Relationship Management: Human connections that survive technological transformation.
Continuous Learning: Adaptation to emerging AI capabilities and market changes.

The Consultant's Choice: Lead the Transformation or Become Obsolete

Every consultant faces the same fundamental choice: embrace AI transformation and develop new value creation models, or resist change and face inevitable displacement.

The consulting industry's $175 billion annual revenue will not disappear—it will redistribute toward organizations and individuals who can deliver superior outcomes through AI enhancement while maintaining human value-add that AI cannot replicate.

Traditional consulting firms that resist AI integration will lose market share to AI-native competitors and AI-enhanced entrepreneurs who can provide better solutions at lower costs with faster Delivery.

Individual consultants who master AI orchestration, develop platform participation strategies, and build communities around AI-enhanced problem-solving will capture market opportunities that large firms cannot address due to organizational constraints.

The transformation is already underway. The only Question is whether you'll lead the change or be displaced by those who embrace it more aggressively.

The consulting transformation reveals a profound truth: when knowledge becomes commoditized, value shifts to something AI cannot replicate—human consciousness. Harnessing this creates an entirely new economy where thoughts become currency.

From my experience implementing enterprise technology solutions for over thirty years, I've learned that successful adaptation requires acknowledging reality rather than hoping it will change. The fact is that AI has commoditized traditional consulting expertise while creating opportunities for new value-creation models.

The consultant's dilemma isn't whether to adapt but how quickly and effectively to embrace the transformation already reshaping our industry.

The Speed Advantage: Individual Consultants vs. Large Firms

The AI revolution creates competitive advantages that favor speed and agility over scale and resources. This transformation enables the Thought Economy by proving that individual expertise, when AI-enhanced, can compete with institutional resources.

Competitive Dynamics:

- **Immediate AI Adoption:** Individual consultants deploy capabilities instantly vs. committee-based corporate evaluation
- **Direct Relationships:** Personal connections enable rapid feedback cycles that large firms cannot match
- **Customized Solutions:** AI enhancement enables personalized service for specific client contexts
- **Continuous Innovation:** Constant experimentation provides advantages large firms cannot replicate

Platform Economy Transition: The consulting transformation demonstrates how professional expertise evolves into platform participation. Individual consultants become thought leaders who contribute frameworks and methodologies to collaborative communities.

Key Takeaways:

- **The $175 Billion Question:** When PhD-level analysis costs $20/month, what justifies premium pricing for human delivery?
- **Three Futures:** Transform (AI partnership), Specialize (narrow expertise), or Disappear (traditional approach)
- **Agility Beats Scale:** Individual AI-enhanced consultants outcompete large traditional firms
- **Platform Economy Evolution:** From selling time to orchestrating intelligence and building collaborative communities

PART IV: THE FUTURE ECONOMY

Now we reveal the economic model that emerges from AI transformation: the Thought Economy. This isn't speculation—it's the inevitable result of AI democratizing technical capabilities while making human consciousness the scarce, valuable resource. These chapters show you how to position yourself to capture value rather than become displaced.

Chapter 11: The Flower Market of Human Consciousness

When Thoughts Become Currency in the AI Economy

The Morning Market Metaphor

Picture this: Five in the morning—pre-dawn darkness. The wholesale flower market is already alive with movement.
Vendors reverse their trucks up to the loading bays, engines still warm from the drive-in from rural farms. The air fills with the scent of fresh blooms—roses, lilies, and chrysanthemums—all harvested just hours ago while most of the city slept. Everything depends on freshness here: peak quality, perfect timing.

By sunrise, the best flowers are gone. What remains loses value rapidly as the day progresses. Tomorrow, the cycle begins again with an entirely new harvest.

The Flower Market is the perfect metaphor for the emerging economy: a marketplace where human thoughts, perspectives, and insights are traded like fresh commodities, with AI systems and businesses as eager buyers seeking the unique perspectives that only human consciousness can provide.

Welcome to the Flower Market of Human Consciousness, where your thoughts have become currency, your experiences hold value, and your unique perspective on the world can become monetized in ways we're only beginning to understand.

The Fresh Flower Market Vision

Each morning brings a fresh harvest of human insights into this new economy. Vendors—individuals with unique perspectives, experiences, and knowledge—arrive at digital marketplaces with their intellectual wares. The "flowers" they're selling are thoughts: creative solutions to business problems, cultural insights that AI cannot access, ethical judgments that require human consciousness, and experiential knowledge that comes only from living in the world.

The buyers are AI systems, businesses, creative projects, and other humans seeking insights they cannot generate independently. They move quickly through the marketplace, selecting the freshest thinking, the unique perspectives, and the insights that sparkle with authenticity and human understanding.

As flower markets prize freshness, the thought economy values originality and timeliness. A business insight that helps a company navigate a challenge is worth more than abstract knowledge. A cultural observation that illuminates emerging trends commands premium pricing. A creative perspective that solves a pressing problem becomes immediately valuable.

And like flower markets, reputation matters. Vendors who consistently deliver high-quality insights, understand what buyers need, and can articulate complex ideas clearly—these thought vendors develop followings and command higher prices for their intellectual harvest.

Why Thoughts Become Valuable

The spatial inversion we explored in Chapter 8 creates geographic freedom that enables global participation in thought markets. The automation we examined in Chapters 7 and 9 eliminates routine cognitive tasks, highlighting the premium value of uniquely human thinking. The consulting transformation detailed in Chapter 10 demonstrates how knowledge becomes commoditized while perspective remains scarce.

But why do thoughts specifically become currency in an AI-saturated economy?

AI Democratizes Technical Capabilities, Not Human Perspective

AI can analyze data, generate reports, create presentations, and efficiently solve structured problems. What AI cannot do is experience the world as a human, understand cultural nuances through lived experience, or generate insights that require consciousness, emotion, and subjective interpretation.

When everyone has access to the same AI capabilities for technical analysis, the differentiating factor becomes the human perspective. The business executive who grew up in rural Montana brings different insights into supply chain challenges than the one who spent decades in urban environments. The healthcare worker who immigrated from another country sees patient care differently than someone born into the local medical system.

These perspectives cannot be automated or replicated. They emerge from the unique combination of genetics, experiences, culture, relationships, and consciousness that defines human existence.

Unique Experiences and Cultural Contexts Remain Irreplaceable

AI training data reflects a broad spectrum of human knowledge and behavior, but it cannot fully capture the unique experiences of individuals in specific locations and moments. In a world where technical capabilities are widely accessible, local knowledge, cultural understanding, and personal insights become highly valuable.

Consider a business expanding into a new market. AI can provide demographic data, economic analysis, and competitive intelligence. However, understanding how local culture affects customer behavior, what unwritten social rules influence business relationships, or how historical events shape current attitudes requires human consciousness to be embedded in that specific cultural context.

Creative Insights and Ethical Judgments Require Human Consciousness

The most valuable human contributions to an AI economy involve creativity, ethics, and judgment—capabilities that emerge from consciousness rather than computation. While AI can generate variations on existing ideas, truly creative insights require the kind of intuitive leaps, pattern recognition, and imaginative synthesis from human consciousness.

Similarly, ethical judgments require understanding context, weighing competing values, and making decisions based on principles beyond algorithmic optimization. As AI systems become more powerful, the need for human ethical oversight and judgment becomes more critical, not less.

The Economic Logic

Information Abundance Creates Attention Scarcity

We live in an era of information abundance. The content created daily is more than anyone could consume in a lifetime. AI amplifies this abundance by generating reports, analyses, and content at an unprecedented scale. In this environment, attention becomes a scarce resource.

Human curation and perspective become differentiators. When everyone has access to the same information, value comes from knowing what to pay attention to, how to interpret data in context, and what insights matter for specific decisions.

Authenticity and Lived Experience Command Premium Pricing

Authenticity has become a premium brand in a world where AI can generate convincing content on almost any topic. Perspectives that come from genuine lived experience, insights that reflect real-world knowledge, and opinions that emerge from actual engagement with problems—these carry weight that AI-generated content cannot match.

Business leaders, investors, and decision-makers increasingly value insights from people who have implemented solutions, faced challenges and learned from real-world experience rather than theoretical analysis.

Platform Economics Enable Global Thought Trading

Digital platforms eliminate geographic barriers to thought trading. A rural farmer with insights into sustainable agriculture can sell those perspectives globally to urban vertical farming companies. A small-town entrepreneur who solved a distribution challenge can monetize that insight for e-commerce businesses worldwide.

Platform economics create network effects where the value of participation increases as more thought vendors and buyers join the marketplace. Quality curation and reputation systems help buyers identify valuable insights, enabling thought vendors to build followings and command higher prices.

Your Platform Vision Genesis

I recognized the thought economy's potential by observing my consulting transformation through AI adoption. As I developed expertise in AI-enhanced business transformation, I realized that the insights I generated—the frameworks, strategies, and solutions emerging from combining human judgment with AI capabilities—had value beyond individual client relationships.

The business transformation insights I developed through decades of technology implementation, enhanced by AI collaboration, represented intellectual property that could benefit multiple organizations simultaneously. Rather than selling time to individual clients, I could package insights for broader distribution.

The pattern recognition skills developed through thirty years of technology consulting—understanding how businesses adapt to change, where resistance emerges, and what strategies succeed—these perspectives become more valuable as AI transformation accelerates and more organizations face similar challenges.

However, the breakthrough insight came from recognizing the parallel between traditional commodity markets and the emerging thought economy. Just as flower markets create organized trading systems for perishable goods, thought markets need organized platforms for intellectual commodities that lose value over time but provide immediate utility to buyers.

Recognition That AI Expertise Has Tradeable Value

My AI expertise itself became a marketable commodity. Organizations struggling with AI adoption pay for insights about implementation strategies, change management approaches, and technical integration challenges. These insights combine technical knowledge, business experience, and human judgment in ways that pure AI analysis cannot replicate.

Understanding That Business Transformation Insights Are Marketable

The frameworks I developed for understanding spatial inversion, automation impact, and workforce transformation have value for organizations across multiple industries. Instead of redeveloping these insights for each client engagement, I could create platform-based access that serves various organizations while generating recurring revenue.

Seeing the Pattern Between Flower Markets and Idea Markets

The connection between commodity and thought trading became clear: both involve perishable goods with maximum fresh value, require quality assessment and reputation systems, and benefit from organized marketplaces that efficiently bring buyers and sellers together.

Platform Concept: Organized Marketplace for Human Consciousness

The platform vision involves creating organized marketplaces where human consciousness—thoughts, insights, perspectives, and experiences—can be traded efficiently between people with valuable intellectual commodities and organizations or individuals needing specific human insight.

The Symbiotic Relationship: Why AI Needs Human Creativity

The thought economy emerges from a fundamental symbiotic relationship between artificial intelligence and human consciousness. Rather than competing for the same functions, AI and human intelligence complement each other in ways that create new forms of value.

Why AI Needs Human Creativity

AI Training Represents Past Human Knowledge and Decisions

Every AI system has devoured all human-generated data representing historical human knowledge, decisions, and behavior. AI can recombine this information sophisticatedly but cannot create genuinely new knowledge or perspectives beyond its training data.

AI Limitation: Cannot Create Truly Novel Perspectives

While AI excels at pattern recognition and can generate variations on existing themes, it cannot create the kind of breakthrough insights that come from human creativity, intuition, and the ability to make connections that transcend logical analysis.

AI Bias Reflects Historical Data Patterns and Limitations

AI systems inherit the biases, limitations, and blind spots present in their training data. Human consciousness provides the external perspective to identify these limitations and generate insights beyond historical patterns.

Human Value: Fresh Thinking, Ethical Judgment, Creative Leaps

Humans contribute fresh thinking that hasn't been incorporated into AI training data, an ethical judgment that considers values beyond optimization, and creative leaps that connect disparate ideas in ways that pure analysis cannot achieve.

The Creative Feedback Loop

The most productive relationship between AI and human consciousness involves a creative feedback loop where each enhances the other's capabilities:

Human-AI Collaboration

- Humans provide creative direction and strategic insight.
- AI handles analysis, research, and content generation.
- Humans add judgment, ethics, and contextual understanding.
- AI amplifies human insights through systematic analysis and broad knowledge access.

Continuous Learning and Adaptation

- Human insights improve AI system performance and accuracy.
- AI capabilities enhance human analytical and creative capacity.
- Both human and AI capabilities evolve through collaborative interaction.
- The combination produces outcomes superior to either human or AI working independently.

Value Creation Through Synthesis

- Human consciousness provides the creative spark and strategic direction.
- AI provides the analytical power and knowledge synthesis
- Together, they create insights and solutions that neither can generate alone.
- The collaborative output becomes the basis for thought trading in platform marketplaces.

New Revenue Streams: From Toothpaste Concepts to Governance Votes

The thought economy creates multiple revenue streams, transforming how human consciousness generates economic value—these range from micro-transactions for simple insights to substantial payments for complex strategic guidance.

Micro-Insights and Daily Observations

Consumer Product Insights

Business insight is as simple as "This toothpaste tube design makes it hard to get the last bit of product out." It becomes valuable when it reaches product designers who can improve customer experience. A platform connecting consumer observations with companies seeking user feedback creates value for observers and businesses.

Process Improvement Suggestions

Workers who notice inefficiencies in business processes, better ways to organize workflows or solutions to recurring problems can monetize these observations through platforms that connect operational insights with management teams seeking improvement opportunities.

Cultural Translation Services

Individuals who understand various cultures can offer insights into how products, services, or marketing strategies will be perceived in different cultural contexts. This knowledge becomes valuable for businesses expanding globally or serving diverse populations.

Strategic Business Insights

Industry Pattern Recognition

Professionals who recognize patterns across their industries—emerging trends, successful strategies, failure modes—can package these insights for sale to organizations facing similar challenges.

Implementation Experience

People who have successfully implemented solutions to business problems can monetize their experience by providing strategic guidance to organizations facing similar challenges. This goes beyond consulting to include structured insight sharing that multiple organizations can access.

Market Analysis and Opportunity Identification

Individuals with deep knowledge of specific markets can provide insights about opportunities, threats, and strategic approaches that AI analysis might miss due to a lack of contextual or experiential learning.

Governance and Decision-Making Participation

Advisory Board Participation

Distributed decision-making platforms enable individuals with relevant expertise to participate in organizational governance through insight contribution rather than traditional board membership.

Crowdsourced Strategic Planning

Organizations can access collective intelligence for strategic planning by creating platforms where multiple experts contribute insights about challenges, opportunities, and strategic approaches.

Ethical Review and Guidance

As AI systems become more powerful, the need for human ethical oversight increases. Platforms connecting organizations with individuals with relevant ethical expertise create new revenue streams for thoughtful analysis of AI deployment decisions.

Creative and Innovation Services

Ideation and Brainstorming

Creative professionals can monetize their ideation capabilities through platforms that connect businesses seeking innovation with individuals who have demonstrated creative problem-solving abilities.

Cross-Industry Innovation Transfer

People who recognize how solutions from one industry might apply to challenges in another sector can monetize these insights through platforms that facilitate innovation transfer.

Future Scenario Development

Individuals skilled at scenario planning and future thinking can provide strategic value to organizations by developing plausible scenarios for how current trends might evolve and what strategic implications might emerge.

Platform Economics: How Thought Trading Markets Function

The thought economy operates according to platform economics principles that create value through network effects, quality curation, and reputation systems.

Supply and Demand Dynamics

Supply: Individual Perspectives, Experiences, and Insights

The supply side consists of individuals with valuable perspectives, experiences, or insights that others find helpful. This includes domain expertise, cultural knowledge, and creative thinking and problem-solving capabilities.

Demand: AI Systems, Businesses, and Creative Projects Needing Human Input

The demand side includes AI systems that need human judgment to improve their outputs, businesses seeking insights they cannot generate internally, and creative projects that benefit from diverse perspectives and experiences.

Pricing: Value-Based on Uniqueness, Accuracy, and Applicability

Pricing reflects the insight's uniqueness, accuracy, relevance to buyer needs, and immediate application to real-world challenges. Rare insights command premium pricing, while more common perspectives trade at commodity rates.

Quality Control: The success of the Community Validation and Expert Review

Platform depends on quality control mechanisms that help buyers identify valuable insights and thought vendors build reputations for reliability and usefulness.

The Curation Value

Not All Thoughts Are Equally Valuable

The democratization of thought trading doesn't mean all perspectives have equal value. Platforms must provide curation mechanisms that help buyers identify insights worth purchasing while helping thought vendors understand what types of contributions are most valued.

Platform Provides Quality Filtering and Validation

Successful platforms develop sophisticated systems for quality assessment that consider the relevance, accuracy, originality, and practical applicability of insights offered by thought vendors.

Expert Reviewers Ensure Insight Quality and Relevance

Platform sustainability requires expert review systems where recognized authorities in specific domains evaluate the quality and relevance of insights before they reach buyers.

Reputation Systems Build Trust and Command Premium Pricing

Thought vendors who consistently provide valuable insights build reputations that enable them to command higher prices and attract more buyers over time.

Global Market Access

Local Expertise Reaching Global Audiences

Platform economics enable individuals with local expertise to serve global markets. For example, a small business owner who solved a particular operational challenge can share that insight with similar businesses worldwide.

Cultural Insights Valuable Across Different Markets:

Understanding specific cultures, languages, or regional markets becomes valuable to organizations operating globally or seeking to expand into new geographic areas.

Language and Regional Knowledge Monetization

Native speakers and cultural insiders can monetize their understanding of language nuances, cultural preferences, and regional business practices through global platform participation.

Cross-pollination of Ideas Across Geographic Boundaries

Platforms facilitates the transfer of successful solutions from one geographic region to another, enabling thought vendors to monetize insights across multiple markets simultaneously.

Geographic Freedom: Live Anywhere, Contribute Globally

The spatial inversion documented in Chapter 7 combines with platform economics to create unprecedented geographic freedom for thought economy participants.

Location Independence for Value Creation

Physical Location is Irrelevant for Thought Contribution.

Unlike traditional employment, which requires presence in specific locations, thought trading can happen from anywhere with internet connectivity. The value comes from the quality of insights rather than geographic proximity to buyers.

Global Audience Access Regardless of Geographic Constraints,

Platform participants can access international markets for their insights without needing to relocate or travel. A rural entrepreneur can sell business insights to urban companies worldwide.

Cultural Diversity as Competitive Advantage

Living in different locations provides competitive advantages through exposure to other cultures, business practices, and problem-solving approaches that create unique perspectives valuable to global audiences.

Time Zone Differences Enabling 24/7 Marketplace Operation

Global platform participation creates a natural 24/7 marketplace operation. Participants in different time zones contribute insights continuously, ensuring buyers always have access to fresh thinking.

The New Digital Nomad Economy

Revenue Generation Independent of Location

Thought economy participation enables revenue generation completely independent of physical location, creating new possibilities for lifestyle design and geographic flexibility.

Participation in the Global Talent Competition and Collaboration

Platform involves competing and collaborating with talent globally rather than just locally, creating opportunities for rapid skill development and income growth.

Cultural Arbitrage Opportunities

Living in locations with lower costs while earning revenue from global markets creates economic opportunities that traditional employment models cannot match.

Community Building Across Geographic Boundaries

Platform participation enables community building with like-minded individuals globally rather than constrained by local networks and opportunities.

The Meta-Moment: You're Living Your Theory

The recursive loop spans 41 years: In 1983, Gödel, Escher, and Bach taught me about strange loops and self-reference. In 1984, I tried to build AI systems that could think about their thinking. In 2025, I wrote a book about the thought economy using thought economy methods. The book that inspired my first AI attempt became the structural template for demonstrating AI's ultimate achievement—the monetization of human consciousness itself.

As I write this chapter about the thought economy, I realize something extraordinary: every conversation I've had with AI while researching and writing this book has perfectly demonstrated the flower market concept.

I brought my raw thoughts—about spatial inversion, robotic restaurants, consulting transformation, and platform economics—to AI systems. Through collaborative dialogue, those thoughts became structured frameworks, business insights, and the content you're reading now.

The Creative Feedback Loop in Action

Each chapter represents the symbiotic relationship between human consciousness and AI capability:

- **Human insight:** Personal experience, pattern recognition, creative connections.
- **AI analysis:** Research synthesis, structural organization, comprehensive development.
- **Combined output:** Frameworks and insights that neither humans nor AI could create independently.

Value Creation Through Synthesis

The book demonstrates how human thoughts become valuable through AI collaboration:

- **Raw thoughts → Structured insights**
- **Individual observations → Connected patterns**

- **Personal experience → Universal frameworks**
- **Conversation → Book content**

The Recursive Loop

The author used this idea to write a book about the thought economy. This recursive relationship illustrates the concept while demonstrating its practical application.

Platform Launch Strategy: You're Already a Founding User

Here's the remarkable truth: if you've engaged with the ideas in this book, thought about their implications for your career and business, or imagined how you might apply these frameworks to your situation—you're already participating in the thought economy.

Your mental engagement with these concepts represents the kind of intellectual value creation platforms will monetize. Your "aha moments," critical thoughts, and creative applications of these frameworks have economic value in a world where human perspective becomes currency.

Early Adopter Advantages

Readers who understand the thought economy concept early have significant advantages:

- **Market timing:** Entering thought markets before they become crowded.
- **Skill development:** Building capabilities in AI collaboration and insight generation.
- **Network effects:** Connecting with other early adopters for collaborative value creation.
- **Platform influence:** Helping shape how thought trading platforms develop.

Community Building Opportunity

This book creates a community of people who understand the concept of thought economy and are positioned to participate in its development. Early readers become founding members of platforms that emerge to serve thought trading markets.

Investment in Your Future

Understanding the thought economy is an investment in your economic future. As traditional employment models face disruption from automation and spatial inversion, thought trading provides alternative revenue streams that leverage uniquely human capabilities.

The flower market of human consciousness isn't a distant future possibility—it's an emerging reality that early participants can help shape while building sustainable economic value from their most uniquely human asset: their consciousness itself.

Revenue Stream Categories

Micro-Insights ($1-$100 per insight):

- Consumer observation: "This app's login process is confusing"
- Process improvement: "This workflow step is redundant"
- Cultural translation: "This marketing message won't work in [specific region]"
- Error identification: "This assumption about customer behavior is incorrect"

Strategic Insights ($100-$5,000 per framework):

- Industry pattern recognition: "This trend indicates market shift toward X"
- Implementation experience: "Here's what actually works when deploying Y"
- Market opportunity identification: "This gap exists between current solutions and customer needs"
- Competitive analysis: "This company's strategy has these specific vulnerabilities"

Governance Participation ($1,000-$10,000 per engagement):

- Advisory board insights for specific decisions
- Ethical review of AI deployment strategies
- Strategic planning contribution for complex challenges
- Innovation ideation for breakthrough thinking

Platform Participation Strategy:

Phase 1: Contribute valuable insights to establish reputation
Phase 2: Develop thought leadership through consistent quality
Phase 3: Build following and command premium pricing
Phase 4: Create frameworks that others use and build upon

Key Takeaways:

- **Fresh Flower Market Metaphor:** Human thoughts are perishable commodities with maximum value when fresh and relevant
- **Economic Logic:** AI democratizes technical capabilities, making human perspective the scarce, valuable resource
- **Revenue Diversification:** Multiple income streams from micro-insights to governance participation
- **Platform Economics:** Network effects increase value as more thought vendors and buyers participate

Chapter 12: The Three-Player Game: Incumbents, AI-Natives, and Entrepreneurs

The Real Competitive Landscape Emerging in the Thought Economy

The thought economy doesn't emerge in a vacuum—it develops within a competitive landscape where traditional incumbents, AI-native companies, and AI-enhanced entrepreneurs compete for market positions. Understanding this three-player game becomes essential for strategic positioning in the thought economy, determining who captures the value created by human consciousness trading platforms.

Most business analysis still frames AI disruption as a simple binary: old companies versus new companies, analog versus digital, traditional versus innovative. This perspective misses the reality of what's happening in markets transformed by AI capabilities.

The competitive landscape involves three distinct player types, each with fundamentally different advantages, limitations, and strategic options. Success in the AI economy requires understanding these dynamics rather than defaulting to outdated competitive frameworks.

The New Competitive Landscape

Beyond Binary Competition

The traditional competitive analysis assumes two-player games: established companies defending a market position against disruptive startups, incumbent consultants competing against technology-enabled alternatives, and existing business models fighting to survive against digital replacements.

This binary thinking worked when technological change followed predictable patterns—when disruption meant faster, cheaper, or more convenient versions of existing solutions. AI transformation operates differently. It doesn't just improve existing capabilities; it creates entirely new categories of value creation that require different competitive strategies.

The three-player game emerges because AI democratizes specific capabilities while creating new scarcity points. Technical implementation becomes accessible to everyone, but human insight, relationship management, and creative problem-solving become more valuable. Platform economics enable collaboration rather than pure competition, and speed advantages favor agile individual actors over resource-rich institutions.

Why Three Players, Not Two

Each player type possesses distinct structural advantages that create unique competitive positioning opportunities:

Traditional Incumbents maintain advantages in established relationships, regulatory compliance, institutional credibility, and resource deployment. They understand industry dynamics, possess decades of client trust, and can invest heavily in transformation initiatives.

AI-Native Companies build competitive advantages around AI capabilities from inception, offering superior analysis and implementation at lower costs than traditional alternatives. They design organizations, processes, and business models optimized for AI-enhanced operations.

AI-enhanced entrepreneurs combine AI capabilities with human judgment, relationship skills, and platform participation to create hybrid value propositions that exceed traditional consulting and pure AI solutions.

Market dynamics favor different players in different situations. Large-scale enterprise implementations favor incumbents with proven track records and implementation capabilities. Standardized analysis and routine problem-solving favor AI-native solutions with superior cost structures. Complex strategic challenges requiring human judgment and relationship management favor AI-enhanced entrepreneurs who combine technological capabilities with personalized service.

Platform economics changes these dynamics by enabling collaboration rather than pure competition. Instead of winner-take-all markets, the thought economy creates opportunities for specialization and partnership across all three player types.

Player 1: Traditional Incumbents - The Advantage of Resources

Strengths That Survive AI Disruption

Large consulting firms, established corporations, and traditional service providers maintain significant advantages that AI transformation doesn't eliminate:

Institutional Relationships: Decades of client relationships create trust and communication channels that AI-native competitors must build from scratch. Enterprise clients often prefer working with established providers for strategic initiatives, regardless of technological capabilities.

Implementation Scale: Large-scale transformation projects require project management capabilities, resource coordination, and risk management that individual entrepreneurs and small AI-native companies cannot match.

Regulatory Navigation: Complex industries require expertise in compliance, regulatory frameworks, and risk management, which comes from institutional experience rather than technological sophistication.

Financial Resources: Major investments in AI capabilities, talent acquisition, and platform development favor organizations with substantial capital resources and the ability to sustain losses during transformation periods.

Brand Credibility: Enterprise clients often require vendor credibility and proven track records for major strategic initiatives. Established brands provide risk mitigation that newer alternatives cannot offer.

The Incumbent Dilemma: Transform or Disappear

However, these advantages come with structural disadvantages that create the Incumbent Dilemma: the characteristics that provide competitive advantages also create barriers to successful AI transformation.

Organizational Inertia:

Large organizations struggle with the rapid decision-making required for AI adoption: committee-based approval processes, risk management protocols, and consensus-building requirements slow adaptation to technological change.

Legacy Infrastructure:

Existing systems, processes, and business models create switching costs and complexity that delay AI integration—the more successful the traditional model, the greater the resistance to transformation.

Cultural Resistance:

Professional service cultures built around human expertise often resist AI augmentation that appears to threaten traditional value propositions and career advancement pathways.

Economic Cannibalization:

AI-enhanced service delivery threatens revenue models built around billable hours, proprietary methodologies, and expertise premiums. Transformation requires abandoning profitable current operations for uncertain future alternatives.

Strategic Options for Incumbents

Traditional incumbents face three strategic options, each with different implications for competitive positioning in the AI economy:

Option 1: Comprehensive Transformation

This is a Fundamental reorganization around AI-enhanced service delivery, platform participation, and collaborative value creation. It requires massive cultural change, technology investment, and business model evolution.

Option 2: Strategic Specialization

Focus on high-value services that require human expertise, relationship management, and complex problem-solving while partnering with AI-native providers for standardized capabilities.

Option 3: Gradual Integration

This approach involves Incremental adoption of AI capabilities while maintaining traditional business models and client relationships. It minimizes disruption but risks a competitive disadvantage against more aggressive adopters.

Most incumbents choose Option 3 by default, which explains why traditional consulting firms continue expanding while individual AI-enhanced entrepreneurs capture increasing market share in strategic advisory services.

Player 2: AI-Native Companies - The Technology Advantage

Built for AI-Enhanced Operations

AI-native companies design organizations, processes, and business models optimized for AI capabilities from inception. Instead of retrofitting AI into existing structures, they build competitive advantages around technological sophistication and operational efficiency.

Technology-First Design:

Every process, workflow, and client interaction incorporates AI capabilities as foundational rather than supplementary. This methodology creates operational efficiencies that traditional competitors cannot match without fundamental reorganization.

Cost Structure Advantages:

AI-enhanced operations enable dramatic cost reductions in analysis, research, document production, and routine problem-solving. These savings translate to competitive pricing that undercuts traditional service providers by 50-80% for comparable outcomes.

Scalability Potential:

AI capabilities scale more efficiently than human expertise, enabling rapid expansion into new markets, service categories, and geographic regions without proportional increases in personnel and overhead costs.

Data Advantage:

Digital-native operations generate data assets that improve AI performance over time, creating competitive moats through machine learning and algorithmic optimization.

The Limitations of Pure AI Solutions

However, AI-native approaches face structural limitations that create opportunities for human-enhanced alternatives:

Relationship Deficit:

Pure AI solutions struggle with complex client relationships, cultural sensitivity, and nuanced communication requirements that characterize high-value strategic engagements.

Creative Limitations:

AI excels at analysis and optimization but struggles with creative problem-solving, innovative strategy development, and breakthrough thinking that requires human insight and intuition.

Context Challenges:

AI systems work best with clearly defined problems and standardized inputs. Complex strategic challenges requiring industry expertise, cultural understanding, and judgment-based decision-making exceed AI capabilities.

Trust Barriers:

Many clients prefer human interaction for strategic decisions, sensitive information sharing, and high-stakes problem-solving, regardless of AI technical capabilities.

Market Positioning for AI-Natives

AI-native companies succeed by focusing on market segments where technological advantages outweigh relationship and creative limitations:

Standardized Analysis:

Financial modeling, market research, competitive intelligence, and data-driven decision support where speed and accuracy matter more than relationship management.

Process Optimization:

Operations improvement, workflow design, and efficiency optimization where systematic analysis produces measurable results.

Technical Implementation:

Technical expertise matters more than strategic guidance in software deployment, system integration, and technology adoption.

Scale Operations:

Large volume, routine projects where cost advantages and processing speed create overwhelming competitive benefits.

The most successful AI-native companies recognize these limitations and develop partnership strategies with human experts for complex strategic work while dominating markets suited to AI capabilities.

Player 3: AI-Enhanced Entrepreneurs - The Wild Card Advantage

The Hybrid Competitive Model

AI-enhanced entrepreneurs represent the most interesting players in the three-way competitive game because they combine advantages from incumbents and AI natives while avoiding many structural limitations that constrain larger organizations.

My consulting practice evolution provides a case study for this hybrid competitive model. Instead of choosing between traditional Expertise and AI capabilities, AI-enhanced entrepreneurs integrate both approaches to create value propositions that exceed pure AI solutions and traditional consulting.

Speed Without Bureaucracy:

Individual entrepreneurs can adopt new AI capabilities immediately, experiment with innovative approaches, and optimize service delivery based on direct client feedback without organizational approval processes or committee-based decision-making.

Relationships Plus Technology:

Personal client relationships combined with AI-enhanced capabilities provide superior outcomes at competitive costs. Clients receive technological sophistication and human attention that larger organizations cannot match.

Customization at Scale:

AI capabilities enable personalized service delivery that addresses specific client needs rather than standardized methodologies designed for broad applications across large client bases.

Platform Participation:

Individual entrepreneurs can participate in collaborative platforms, contribute to knowledge-sharing communities, and develop new revenue streams without corporate constraints or competing priorities.

My Strategic Evolution: From Consultant to Platform Participant

My consulting practice transformation demonstrates how individual knowledge workers can navigate AI disruption by embracing rather than resisting technological change while developing new value creation models that survive AI commoditization of traditional expertise.

Phase 1: Traditional Expertise (1990s-2010s)

Built competitive advantages around specialized knowledge in enterprise systems, Office 365 deployment, and POS system integration. Value creation came through knowledge scarcity and implementation experience.

Phase 2: AI-Enhanced Delivery (2020s)

Integrated AI capabilities to improve analysis speed, solution quality, and client outcomes. Maintained traditional consulting relationships while enhancing service delivery through AI augmentation.

Phase 3: Platform Participation (Current)

Developing communities around AI-enhanced business transformation, contributing to shared knowledge platforms, and creating value through collaborative problem-solving rather than individual consulting engagements.

Phase 4: Platform Leadership (Future)

Contributing to platform governance, methodology development, and community building while maintaining specialized expertise in spatial inversion, automation strategies, and business transformation.

This evolution demonstrates how individual consultants can navigate AI disruption by embracing rather than resisting technological change while developing new value-creation models that survive AI commoditization of traditional expertise.

The Speed Advantage: Why Agility Beats Scale

The most significant competitive advantage for AI-enhanced entrepreneurs is speed and flexibility rather than scale and resources. While large firms debate AI strategy through committee processes and navigate organizational resistance, individual entrepreneurs immediately deploy new capabilities and adjust service delivery based on client feedback and technological advancement.

Immediate AI Tool Adoption:

While large firms evaluate AI strategies through committee processes, I deploy new capabilities immediately and optimize based on client results. The latest AI tools become available immediately rather than after months of vendor evaluation and corporate approval.

Direct Client Relationships:

Personal connections with decision-makers enable rapid communication and feedback cycles that large firms cannot match through account management structures. The client must be addressed immediately rather than through relationship managers and project teams.

Customized Solutions:

AI enhancement enables personalized service delivery that addresses specific client contexts rather than standardized methodologies designed for broad applications. Each engagement becomes customized for maximum client value.

Continuous Innovation:

Constant experimentation with new AI capabilities and integration strategies provides competitive advantages that large firms cannot match due to organizational constraints and risk management requirements.

The result is superior client outcomes at lower costs with faster delivery times—a competitive advantage that large consulting firms cannot replicate without fundamental organizational transformation.

Market Dynamics and Strategic Positioning

The Innovation Diffusion Timeline

Understanding market evolution requires recognizing how AI capabilities diffuse across different player types at different speeds:

2023-2024: Early Adopter Advantage

AI-enhanced entrepreneurs and small AI-native companies gain competitive advantages through rapid AI adoption while incumbents evaluate strategies and plan transformation initiatives.

2025-2026: Scale Response

Established companies invest heavily in AI capabilities, pursue acquisition strategies, and respond to competition. They drive market consolidation by scaling successful models for broader implementation.

2027-2028: Platform Emergence

Collaborative platforms enable specialization and partnership across player types, and competition shifts from pure displacement to complementary value creation.

2029+: Ecosystem Maturation

Stable competitive dynamics emerge with each player type optimized for market segments and customer needs.
Strategic positioning requires understanding where your organization or practice fits within this timeline and evolution pattern.

The Strategic Options Matrix

Each player type faces different strategic options based on their structural advantages and limitations:

Incumbents:

Transform completely, specialize strategically, or integrate gradually while risking competitive displacement.

AI-Natives:

Expand into relationship-dependent services, partner with human experts, or dominate technology-suited market segments.

Entrepreneurs:

Scale through platform participation, specialize in complex strategic work, or develop hybrid service models that combine AI capabilities with human expertise.

The most successful strategic positioning combines understanding your structural advantages with a realistic assessment of market dynamics and competitive responses.

Platform Economy Solutions: Collaboration Over Competition

The thought economy creates opportunities for collaboration rather than pure competitive displacement. Platform economics enable different player types to contribute complementary capabilities rather than competing directly for the same market segments.

Incumbents

Contribute established relationships, implementation capabilities, and institutional credibility to platform communities.

AI-Natives

Provide technological infrastructure, analytical capabilities, and cost-efficient processing for platform operations.

Entrepreneurs

Offer specialized expertise, creative problem-solving, and personalized relationship management that enhances the platform value of all participants.

This collaborative model creates larger market opportunities than traditional competitive approaches because it combines the unique advantages of each player type rather than forcing zero-sum competition.

Competitive Intelligence for Strategic Positioning

Understanding Your Competitive Position

Strategic positioning in the three-player game requires an honest assessment of your structural advantages and limitations:

If you're part of a traditional incumbent, Focus on relationships, implementation capabilities, and regulatory expertise while developing AI enhancement strategies that improve rather than replace human capabilities.

If you're building an AI-native company, Concentrate on market segments where technological advantages matter more than relationship management while developing partnership strategies for complex strategic work.

If you're an individual entrepreneur, Embrace AI enhancement immediately while building platform participation strategies and relationship advantages that larger organizations cannot replicate.

Market Timing Strategies

Different strategic moves work better at various phases of market evolution:

Early Phase (Current):

AI enhancement provides competitive advantages for individual entrepreneurs and small companies willing to experiment with new capabilities.

Growth Phase (2025-2026):

As market consolidation accelerates, platform participation becomes essential for sustainable competitive positioning.

Maturation Phase (2027+):

As ecosystem dynamics stabilize, specialization and collaboration will become more critical than competitive displacement.

Understanding market timing helps determine when to make strategic investments and how aggressively to pursue transformation initiatives.

Platform Participation as a Competitive Advantage

The most successful strategic positioning treats platform participation as a competitive advantage rather than an alternative to traditional competition. Instead of competing in traditional markets or participating in platform economies, the most effective approach combines both strategies.

Platform participation provides:

- **Revenue Diversification** beyond traditional client relationships.
- **Knowledge Sharing** that improves service delivery capabilities.
- **Community Building** that creates referral networks and collaborative opportunities.
- **Market Intelligence** about emerging trends and competitive dynamics.

Traditional competitive advantages provide:

- **Client Relationships** that generate platform participation opportunities.
- **Specialized expertise** that contributes value to platform communities.
- **Implementation Experience** that enhances platform-mediated service delivery.
- **Market Position** that provides credibility for platform leadership roles.

The combination creates sustainable competitive advantages that exceed either approach independently.

Strategic Positioning Matrix

Player Type	Structural Advantages	Optimal Strategy	Thought Economy Role
Incumbents	Relationships, Resources, Credibility	Strategic Specialization	Platform Infrastructure
AI-Natives	Technology-First Design, Cost Structure	Market Segment Domination	Analytical Engine
Entrepreneurs	Speed, Agility, Personal Relationships	AI-Enhanced Hybrid Model	Platform Leadership

The Collaboration Opportunity

The Thought Economy creates win-win scenarios where each player type contributes complementary capabilities:

Incumbents provide:

Established relationships, implementation scale, institutional credibility

AI-Natives provide:

Technical infrastructure, analytical capabilities, cost-efficient processing
Entrepreneurs provide: Creative problem-solving, specialized expertise, personalized relationships

Platform Value Creation:

Combined capabilities exceed what any single player type can achieve independently.

Key Takeaways:

- **Beyond Binary Competition:** Three distinct player types with unique advantages create collaboration opportunities
- **Speed vs. Scale:** AI transformation favors agility over resources, enabling individual entrepreneurs to compete with institutions

- **Platform Economics:** Collaboration rather than pure competition creates larger market opportunities
- **Strategic Positioning:** Success requires understanding your structural advantages and choosing appropriate competitive strategy

Chapter 13: The Invisible Divide: AI-Literate vs. Traditional Workforce

The Most Dangerous Competitive Gap of Our Time

Organizations worldwide are investing billions in AI initiatives, yet a significant divide is emerging that could transform entire industries. This divide isn't just between companies that adopt AI and those that don't; it involves the invisible transformation of the workforce within organizations. This shift creates competitive advantages and disadvantages that can grow exponentially over time.

In my consulting work across enterprise transformations—from O365 migrations to UiPath automation implementations, plus developing custom business solutions that address specific pain points using LLMs for research and rapid prototyping to build actual working solutions and proof-of-concept systems—I witness this divide daily. Most organizations are taking a cautious, task-specific approach—investing in AI applications that solve discrete problems like invoice reading, document processing, or customer service chatbots. They're buying AI solutions rather than building AI capabilities.

I observed this pattern firsthand at a recent UiPath event in Sydney: strong corporate turnout from companies in various stages of bringing on UiPath. I am particularly excited about the new Agent functionality. Businesses are actively seeking automation through applications like UiPath, Power Apps, or n8n low-code platforms, but they're approaching these as workflow solutions rather than AI literacy development opportunities.

Meanwhile, AI-first companies face a different challenge: learning to upskill and motivate their existing teams to become AI-efficient. This transformation process is surprisingly complex because it requires fundamental mindset shifts. Experienced programmers and subject matter experts (SMEs) who have developed their toolbox of methods over the years find AI amplification

unsettling—it's like undoing your seatbelt in a car and then driving faster. There's an uncomfortable feeling that it's not safe until you can get your conceptual "seatbelt" back on.
The challenge isn't teaching new skills to new people; it's equipping domain experts with appropriate AI know-how to become AI-efficient while maintaining their expertise value. Effectively managing this requires careful navigation between utilizing their extensive knowledge and adapting their problem-solving methods.

This results in two different patterns of workforce evolution: traditional companies that adopt AI tools in isolation and AI-first companies that strive for comprehensive team transformation. These companies may not realize that their workforce has quietly developed into three distinct tiers based on AI tool usage and differences in strategic thinking and problem-solving approaches.

The statistics reveal the magnitude of this hidden crisis: While 78% of companies globally have adopted AI in at least one business function, 22% of organizations have no AI implementation whatsoever. Among those adopting AI, most % implement task-specific applications—56% use AI for customer service, 51% for cybersecurity, and 40% for inventory management. Companies typically deploy AI across an average of three different business functions, treating AI as a collection of tools rather than a strategic capability.

In stark contrast, only an estimated 5-10% of organizations are investing in comprehensive AI literacy development that transforms how their workforce thinks and operates. Among those attempting true AI transformation, research indicates that only 24% of business leaders demonstrate genuine AI fluency—the ability to think strategically about AI capabilities and limitations rather than deploy AI solutions.

The gaps create a three-tier market reality: 22% with no AI, 68% with task-specific AI tools, and only 5-10% building strategic AI literacy. Organizations in the middle tier develop dangerous false confidence, believing they're "doing AI" because they've purchased AI solutions while remaining unaware of the exponential productivity gaps developing within their workforce. Most critically, traditional performance metrics fail to capture AI literacy gaps, creating the "invisible crisis": competitive disadvantages that develop gradually, compound over time, and remain undetected until they become irreversible.

The Hidden Workforce Evolution

Why This Transformation Is Invisible

Conventional organizational measurement systems do not effectively identify gaps in AI literacy. Performance reviews evaluate task completion rather than strategic thinking evolution. Productivity metrics measure outputs rather than capability transformation. Management often remains unaware of employee AI mastery levels because the work gets done—just not as efficiently or innovatively as possible.

What makes this particularly dangerous is the gradual nature of divergence. AI literacy gaps don't appear overnight as dramatic performance failures. Instead, they compound monthly as AI-literate employees become exponentially more productive while traditional workers maintain baseline performance levels.

In my daily interactions across various client organizations, I regularly encounter scenarios where individual employees are quietly delivering work that should require entire teams. At the same time, their colleagues struggle with basic automation concepts. I'm also seeing people push the use of AI to get work done, and businesses are slowly discovering these productivity disparities, then confronting themselves with addressing loose ends without still seeing the big picture of what's happening.

Leadership often discovers these productivity gaps only when competitors deliver capabilities their teams consider "impossible" or when talent acquisition becomes difficult because top performers expect AI-enhanced work environments.

The Acceleration Factor

Unlike traditional skill gaps that develop linearly over the years, AI literacy gaps compound exponentially because AI capabilities improve continuously. An employee who masters AI orchestration in January possesses fundamentally different capabilities by December—not just because of personal skill development but also because the AI tools become more sophisticated.

I've witnessed this acceleration firsthand from my experience developing POS systems and other products. I can prototype concepts in minutes using strategic AI collaboration but then spent four months learning, vibing, and discovering the pitfalls of what's needed to make an AI product ready for the rapidly changing landscape of how AI will fit in and outperform competitors.

Like this conversation, where we've evolved Chapter 13 and spun off two article ideas in the last two hours, the same acceleration happens when crafting and vibing with LLMs doing the cognitive heavy lifting. The LLM frees up mental capacity and accelerates thinking to explore wider possibilities: how an app should perform and what "nice-to-have" functionality becomes possible. But here's the critical risk: **you can bring in a "nice-to-have" feature in 5 minutes without proper documentation or architectural thinking**.

The organizations and individuals who understand this distinction—between rapid capability expansion and sustainable development practices—build lasting competitive advantages while others create impressive prototypes that become unmaintainable technical debt.

The Three-Tiered Workforce Reality

Based on my observations across dozens of organizations, workforce AI literacy has evolved into three distinct tiers, though percentages vary significantly by industry and are rapidly shifting:

Tier 1: AI-Native Employees (5-15%)

These individuals represent the future workforce operating within current organizations. They don't just use AI tools—they think of AI as a natural extension of their cognitive capabilities.

Characteristics of AI-Native Workers:

Multi-AI Orchestration:

They understand how to leverage different AI systems for their strengths. For example, they use Gemini's 1M token context for system architecture and documentation and then transfer focused components to Claude for detailed implementation. They've moved beyond single-tool dependence to strategic AI ecosystem management.

Strategic AI Thinking:

They understand current AI capabilities and limitations while recognizing that AI represents an entirely different category of technological transformation. Unlike the internet or cloud computing—which were one-time infrastructure shifts—AI is a continuously evolving tech platform that will never stop advancing.

They can assess which AI-generated solutions are viable versus which ones look impressive but would fail in real-world implementation. Most critically, they understand that we're now in a hyper-transformation stage where current business models and systems will fail, not built with continuous AI evolution in mind. Successful organizations must fundamentally reimagine themselves as AI companies that happen to sell X & Y rather than traditional companies that use AI tools.

AI-Native workers grasp the transformational nature of Software 3.0—a software structure built to be transformational, unlike any previous computing paradigm. Once you have a transformational structure with AI at its heart, it enables real-time self-improvement cycles that compound exponentially. Companies that heard about AI during the initial hype cycle are now discovering this isn't a one-time transformation but an ongoing hyper-evolution that requires fundamental business model redesign.

Sustainable Development Practices:

Unlike those who copy-paste AI outputs, they maintain proper documentation, consider technical debt implications, and build maintainable systems as AI capabilities evolve.

Platform Economy Participation:

They contribute to AI communities, develop thought leadership, and create value streams that extend beyond traditional employment through knowledge sharing and collaborative problem-solving.

In my client organizations, these employees drive disproportionate value creation. They complete projects that would require entire teams using traditional approaches. They identify automation opportunities that others miss entirely. Most importantly, they serve as force multipliers for organizational AI transformation.

Tier 2: AI-Enhanced Workers (30-50%)

This group represents the organizational majority: employees who use AI tools effectively but haven't developed the strategic thinking frameworks that enable true AI-native capabilities.

Characteristics of AI-Enhanced Workers:

Tool Proficiency: They master specific AI applications and follow established AI workflows competently, often achieving significant productivity improvements through guided implementations.

Prescribed Usage:

They benefit from AI enhancement primarily through training programs and established procedures rather than independent innovation or multi-system orchestration.

Context Management Challenges:

They often fall into the trap of starting new AI conversations rather than maintaining strategic context across complex projects, leading to fragmented solutions and repeated work.

Limited Technical Debt Awareness:

They focus on immediate productivity gains without considering long-term maintenance, sustainability, or integration challenges.

These employees provide organizational stability during AI transformation and demonstrate that AI enhancement works. However, they also represent significant missed opportunities, as many could transition to Tier 1 capabilities with proper strategic thinking development.

Tier 3: AI-Resistant/Unaware (35-60%)

This group creates the most significant organizational risk because they often include key knowledge holders, experienced professionals, and domain experts whose expertise becomes increasingly isolated from AI-enhanced organizational capabilities.

Critical Risk Factors:

Knowledge Isolation:

Critical institutional knowledge remains trapped in traditional workflows that resist AI integration, creating organizational brittleness.

Productivity Divergence:

Performance gaps between AI-enhanced and traditional workers create internal friction and competitive disadvantages that compound over time.

Technical Debt Multipliers:

When forced to adopt AI solutions, they often create unsustainable implementations that require extensive future remediation.

Leadership Constraints:

The most dangerous scenario occurs when Tier 3 employees occupy leadership positions that control AI strategy and investment decisions, constraining organizational transformation regardless of available resources.

The Great Equalizer Phenomenon

The most striking competitive implication involves the "Great Equalizer Phenomenon"—AI literacy enables small teams and individual entrepreneurs to compete directly with large organizational departments.

Real-World Evidence

In my consulting practice, I regularly outperform consulting teams of 5-10 people because AI enhancement enables analysis, strategy development, and implementation planning that would traditionally require multiple specialists. The competitive advantage comes from AI literacy and strategic orchestration rather than organizational resources.

From my product development experience, solo developers using multi-AI workflows deliver applications faster and with better documentation than traditional 5-person development teams. The key isn't just using AI tools—it's understanding how to orchestrate multiple AI systems while maintaining sustainable development practices.

Which creates unprecedented competitive dynamics:

- Solo entrepreneurs matching enterprise capabilities through strategic AI orchestration
- Small teams delivering outcomes that previously required department-level resources
- Geographic arbitrage as AI-literate individuals provide global services from any location
- Platform economy participation enables individual specialists to access markets previously controlled by large organizations.

The Technical Architecture Advantage

Traditional competitive moats based on scale, resources, and organizational capabilities become less relevant when AI literacy enables individuals to deliver enterprise-level outcomes. However, **sustainable competitive advantage requires more than rapid AI adoption**—it demands an understanding of the following:

System Architecture:
How different AI tools integrate into coherent Workflows.

Documentation Practices:
Maintaining context and knowledge transfer across AI-assisted projects.

Technical Debt Management:
Building solutions that remain maintainable as AI capabilities evolve.

Quality Assurance:
Ensuring AI-generated outputs meet professional standards.

Strategic Implications for Organizations

The Business Decision Threshold

Every organization eventually reaches the "business decision threshold"—recognizing that traditional methods have reached diminishing returns and AI transformation becomes essential for competitiveness.

Critical threshold indicators:

- Competitors delivering capabilities your team considers "impossible."

- Rising customer expectations that current processes can't meet efficiently.
- Increasing difficulty recruiting top talent who expect AI-enhanced work environments.
- Diminishing returns from incremental improvements to existing processes.
- Individual consultants outperform your internal teams on complex projects.

Organizations recognizing this threshold early and investing aggressively in sustainable AI literacy development gain cumulative advantages. Those who realize too late attempt to bridge impossible capability gaps through traditional training approaches or expensive vendor solutions.

The Enterprise Dilemma: Control vs. Speed

Large organizations face a strategic choice between three approaches:

The "Safe" Vendor Path

Involves Waiting for packaged enterprise solutions, maintaining support contracts, and avoiding technical debt, but accepting 2-3 year delays behind first movers.

The "Risky" Internal Path:

Build custom AI capabilities rapidly, accepting technical debt and maintenance responsibility but risking unsustainable systems.

The "Strategic" AI-Native Path:

Develop AI literacy as a core competency, build sustainable practices, and create competitive moats through orchestration capabilities rather than just tool adoption.

The 90-Day Sustainable AI Transformation Program

Based on my experience helping organizations navigate this divide, the most effective approach involves focused transformation programs that balance immediate productivity gains with long-term strategic capability development.

Phase 1: Assessment and Foundation Building (Days 1-30)

- AI Literacy Assessment using real-world scenarios rather than tool checklists.
- Multi-AI Orchestration Training focusing on strategic tool selection
- Sustainable Development Practices emphasizing documentation and maintenance.
- Champion Identification recognizes existing AI-native employees as transformation leaders.

Phase 2: Strategic Capability Development (Days 31-60)

- Advanced AI Thinking frameworks for understanding capabilities and limitations.
- Technical Debt Prevention strategies for long-term system sustainability.
- Cross-platform integration connecting different AI systems for organizational synergy.
- Quality Assurance protocols ensuring professional standards for AI-assisted work.

Phase 3: Organizational Integration and Scaling (Days 61-90)

- The governance framework covers usage policies, quality standards, and strategic oversight.
- Knowledge Management systems that capture AI-enhanced insights and methodologies.
- Competitive Positioning strategies leveraging AI literacy for market differentiation.
- Platform Economy Participation exploring external collaboration and revenue opportunities.

Personal and Professional Positioning Strategies

Assessing Your Current AI Literacy Tier

Understanding your position within this framework enables strategic decisions about skill development and career positioning:

AI-Native Indicators:

- You naturally orchestrate multiple AI systems based on their specific strengths.
- You maintain proper documentation and consider technical debt in AI-assisted projects.
- You participate in AI communities and contribute insights beyond your immediate role.
- You can assess AI-generated solutions for real-world viability and sustainability.

AI-Enhanced Indicators:

- You use AI tools effectively for specific tasks but primarily follow established procedures.
- You achieve significant productivity improvements but haven't developed strategic frameworks.
- You want to advance beyond basic tool usage but need guidance on sustainable practices.
- You recognize AI value but struggle with context management across complex projects.

Traditional Worker Indicators:

- You avoid AI tools due to complexity, skepticism, or lack of training opportunities.
- You rely primarily on traditional methods, even when AI alternatives could provide advantages.
- You feel uncertain about AI implications for your role and the industry's future.
- You need foundational AI literacy development before advancing to strategic applications.
- **Your organization is geo-fenced around customer data privacy regulations** and unable to deploy or understand infrastructure for localized LLMs, creating fundamental competitive disadvantages.

The Geopolitical AI Divide:

This final indicator reveals a critical but overlooked dimension: **regulatory geo-fencing creates AI-have and AI-have-not nations**. In Australia, the Privacy Act 1988 and Australian Privacy Principles (APPs) create significant barriers for organizations with annual turnover over AUD$3 million. APP 8 requires organizations to ensure overseas recipients don't breach APPs, limiting cloud-based AI usage for sensitive data.

Yet there is no official count of "local LLM data centers" in Australia, creating an infrastructure gap that puts Australian organizations at a severe disadvantage compared to AI-first countries like the UAE, which are building national AI infrastructure strategies.

This creates a dangerous competitive reality:

- Australian companies remain geo-fenced by privacy regulations.
- AI-first governments and countries develop unrestricted AI capabilities.
- Organizations avoiding cloud-based AI due to compliance concerns fall further behind.
- The lack of local AI infrastructure forces a choice between compliance and competitiveness.

Organizations stuck in traditional approaches due to regulatory constraints risk becoming competitively obsolete as AI-first companies and countries develop capabilities they cannot legally or practically access.

Strategic Development Pathways

For Traditional Workers:

Focus on foundational AI literacy through structured learning that provides immediate practical applications. Start with AI tools for everyday tasks, then gradually develop an understanding of strategic implications and sustainable practices.

For AI-Enhanced Workers:

Transition to strategic AI thinking by learning multi-system orchestration, participating in AI communities, and developing frameworks for sustainable AI-assisted development.

For AI-Native Workers:

Leverage your capabilities for platform economy participation, thought leadership, and organizational transformation consulting. Your expertise becomes valuable for training, strategic guidance, and community-building opportunities.

The Urgency Imperative

The Compounding Timeline

The invisible divide becomes more dangerous over time because AI literacy gaps compound exponentially rather than develop linearly. Organizations that delay sustainable AI literacy development face complex catch-up challenges as AI capabilities advance and competitive advantages solidify.

The acceleration timeline means that decisions made in the next 12-18 months determine competitive positioning for the next decade. Organizations that build genuine AI literacy now—not just tool adoption—gain cumulative advantages that become impossible for traditional competitors to match through resource investments alone.

Market Selection Pressure

Customer expectations evolve rapidly as AI-enhanced alternatives become available. Markets increasingly select AI-enabled capabilities regardless of organizational preferences for traditional approaches, creating what economists call "selection pressure"—environmental forces that favor specific characteristics over others.

In business markets, this pressure increasingly favors AI literacy over traditional expertise, speed over deliberation, and innovation over established practices. Organizations and individuals who resist this selection pressure compete for shrinking market segments while AI-enhanced alternatives capture expanding opportunities.

Conclusion: The Ecosystem Transformation

The invisible divide between AI-literate and traditional workers reveals why many AI initiatives fail despite significant investments. Organizations focus on tool deployment rather than literacy development, measure adoption rather than capability transformation, and underestimate the cultural and strategic changes required for sustainable AI integration.

The fundamental insight is that the entire business ecosystem needs to change. AI enables rapid prototyping and concept development, transforming how we approach product development, strategic planning, and competitive positioning. Organizations that recognize this ecosystem transformation and invest in sustainable AI literacy development position themselves for long-term competitive advantage.

The choice is no longer whether to adopt AI but whether to build the strategic thinking capabilities that enable sustainable AI advantage. In this transformation, individual AI literacy becomes organizational competitive positioning, and platform economy participation becomes essential for accessing the thought economy opportunities that define future business success.

The invisible divide is becoming visible. Which side will you choose, and how quickly will you act on that choice?

The AI Literacy Assessment Tool

Tier 1: AI-Native Employees (5-15%)

- I orchestrate multiple AI systems based on their specific strengths
- I maintain proper documentation and consider technical debt in AI projects
- I participate in AI communities and contribute insights beyond my immediate role
- I can assess AI-generated solutions for real-world viability and sustainability
- I understand we're in a hyper-transformation stage requiring continuous adaptation

Tier 2: AI-Enhanced Workers (30-50%)

- I use AI tools effectively for specific tasks with significant productivity improvements
- I follow established AI workflows and achieve measurable efficiency gains
- I understand AI benefits but primarily through guided implementation
- I need support for complex context management across projects
- I focus on immediate gains without considering long-term technical implications

Tier 3: AI-Resistant/Unaware (35-60%)

- I avoid AI tools due to complexity, skepticism, or lack of training
- I rely primarily on traditional methods even when AI alternatives exist
- I feel uncertain about AI implications for my role and industry
- My organization has regulatory constraints limiting AI deployment
- I need foundational AI literacy development before advancing to strategic applications

Development Pathways by Tier:

Tier 3 → Tier 2 (3-6 months):

- Foundational AI literacy through structured learning
- Basic tool familiarity with everyday applications

- Understanding of AI capabilities and limitations
- Development of strategic thinking about AI implications

Tier 2 → Tier 1 (6-12 months):

- Multi-system orchestration and strategic AI thinking
- Platform community participation and contribution
- Sustainable development practices and technical debt awareness
- Leadership in organizational AI transformation

Key Takeaways:

- **Three-Tier Reality:** Organizations unknowingly develop AI-native (5-15%), AI-enhanced (30-50%), and traditional (35-60%) workforce segments
- **Exponential Divergence:** AI literacy gaps compound monthly as capabilities advance and early adopters gain cumulative advantages
- **Invisible Crisis:** Traditional performance metrics fail to capture AI literacy gaps until competitive disadvantages become irreversible
- **Urgency Imperative:** Decisions made in next 12-18 months determine competitive positioning for the next decade

PART V: YOUR TRANSFORMATION

This final section provides your complete action plan for participating in the Thought Economy. We move from understanding to implementation, giving you the specific frameworks and strategies to build value from your consciousness in partnership with AI.

Chapter 14: The AI Communication Revolution: Your Business Voice Is Already Changing

AI is gradually and invisibly infiltrating business correspondence.

The sophistication spike, homogenization effect, and complexity explosion reveal transformation patterns most leaders haven't recognized.

You understand the future economy, the competitive landscape, and the workforce transformation. Now, it's time to focus on your transformation. Let's start by discussing how AI is already changing your business communications.

In the silent, unseen currents of digital transformation, something extraordinary happens to business correspondence—and most leaders are oblivious to it.

After spending decades delivering technology solutions—from mechanical machines to AI—for businesses, I've developed an eye for patterns in how companies communicate. Recently, I conducted an analysis that revealed something that should make every business leader pause: AI isn't just changing how we work—it's already changing how we think, write, and make decisions, often without us realizing it.

The evidence isn't in the ChatGPT headlines. It hides in your company's emails, proposals, and internal communications.

The Silent Revolution

What's Happening Without Your Awareness

AI is gradually and invisibly infiltrating business correspondence. Most leaders remain unaware of the communication transformation within their organizations despite being responsible for strategic decisions based on this changing information landscape.

My unique perspective as a technology implementer and communication observer provides pattern recognition across multiple industries and organizational contexts. I've analyzed business correspondence spanning different timeframes, organizational sizes, and industries to identify systematic changes that reveal AI adoption.

AI correspondence affects every professional, not just technology workers. The scope of transformation includes:

Email Communication:

Style sophistication, response quality, and strategic thinking complexity have dramatically improved across organizations that historically struggled with professional correspondence.

Proposal Writing:

Research depth, presentation quality, and persuasion effectiveness show sudden enhancement that exceeds normal skill development timelines.

Strategic Documents:

Analysis complexity, framework sophistication, and conceptual articulation demonstrate capabilities that individual authors never previously displayed.

Internal Communications:

Consistency, professionalism, and clarity improve immediately rather than gradually, suggesting external assistance rather than skill development.

Why This Matters Now

Communication authenticity is becoming a competitive differentiator in markets where AI-generated content creates opportunities and vulnerabilities. Organizations that understand and strategically manage AI communication integration gain advantages over those experiencing unconscious transformation.

AI dependency creates organizational vulnerabilities that most leaders don't recognize until critical communications fail or competitive disadvantages emerge. Understanding this change enables strategic advantage rather than reactive crisis management.

Platform economy participation requires high-quality human communication skills that complement rather than compete with AI capabilities. The thought economy we explored earlier depends on authentic human insight enhanced by, not replaced by, AI capabilities.

The Three Transformation Patterns

Understanding how AI transforms business communication requires recognizing three distinct patterns that emerge consistently across organizations. These patterns reveal both the opportunities and risks of AI integration.

Pattern 1: The Sophistication Spike

Organizations that have historically struggled with clear, professional communication suddenly produce polished, eloquent correspondence. The transformation isn't gradual—it's dramatic and immediate, creating quality improvements that exceed normal skill development timelines.

Before AI Adoption Examples:

"We need to discuss the issues with the current team and figure out what to do next."

- Simple sentence structure, basic vocabulary, unclear objectives

After AI Adoption Examples:

"We should convene a strategic assessment to evaluate system inefficiencies and develop a comprehensive remediation roadmap."

- Complex sentence structure, professional vocabulary, specific action orientation.

Language complexity and professional tone enhancement occur suddenly rather than through gradual skill development. Teams that previously struggled with meeting minutes, project updates, and client communications begin producing documentation that rivals professional consulting standards.

I've observed this pattern across dozens of organizations: accounting firms suddenly writing with marketing sophistication, manufacturing companies producing strategic analysis comparable to management consulting, and small businesses developing proposal quality that matches enterprise standards.

Pattern 2: The Homogenization Effect

Distinct writing voices within organizations converge into remarkably similar styles. A standardized "professional AI tone" that spans departments and hierarchy levels replaces individual personalities in written communication.

Organizations lose communication diversity as AI systems impose consistent style, vocabulary, and structural approaches across authors. Email threads become difficult to attribute to specific individuals because the writing patterns become indistinguishable.

AI communication creates both benefits and risks. Communication consistency improves organizational professionalism and reduces misunderstandings. However, losing individual voice eliminates personality-based relationship building and makes it difficult to assess individual communication competency.

More concerning, management is beginning to decide about communication skills and strategic thinking capabilities based on AI-enhanced rather than authentic individual performance.

Pattern 3: The Complexity Explosion

Due to convoluted language, simple concepts become unnecessarily complex, implying that AI assistance is used not only for efficiency but also as a crutch for perceived professionalism.

Communications that could be conveyed often opt for elaborate language structures, extensive vocabulary, and sophisticated frameworks, even when clarity would better serve the audience. This trend suggests that AI prioritizes perceived sophistication over effective communication.
Examples include:

- Routine status updates carry a strategic assessment structure.
- Simple requests entwined in complex analytical frameworks.
- Basic information wrapped up using advanced business terminology.

The complexity explosion reveals a reliance on AI rather than an enhancement of it—users depend on AI to make their communications appear more professional instead of using it to improve their effectiveness.

The Power Shift Dynamics

These communication patterns create profound changes in organizational hierarchy and competitive advantage that most leaders haven't recognized.

When AI Speed Outpaces Experience

The most significant change isn't in communication quality but organizational power dynamics. Junior staff equipped with AI can now produce strategic analysis and respond to complex scenarios faster than seasoned managers can process the questions.

I've experienced this personally—using AI to formulate comprehensive responses within minutes while colleagues spend hours deliberating. The information advantage is so dramatic that I've learned to delay my responses to avoid deliberately revealing the disparity.
Utilizing AI creates a new workplace hierarchy based on AI fluency rather than experience:

- AI-equipped staff can access current data, analyze market trends, and deliver sophisticated responses in real time.

- Traditional managers rely on institutional knowledge but operate at human thinking speed.
- The strategic gap widens as AI users gain competitive advantages and often feel compelled to conceal them.

The Psychological Advantage

I've begun using AI to analyze incoming emails for writer profiling, which involves understanding communication styles, decision-making patterns, and psychological approaches. This approach allows me to tailor my responses strategically rather than just replying to content. It's like having a behavioral analyst for every business interaction.

The implications extend beyond individual performance. Organizations with AI-fluent teams can respond quickly to market changes, client requests, and strategic challenges. Yet most businesses remain unaware of this shift within their communications infrastructure.

AI-fluent teams demonstrate:

Rapid Response Capability: Complex analysis and strategic recommendations delivered within hours rather than days.

Enhanced Analytical Sophistication: Insights that combine multiple data sources, industry patterns, and strategic frameworks.

Competitive Advantage: Superior client service and internal coordination through enhanced communication quality.
Most businesses remain unaware of internal AI usage transformation, creating opportunities for AI-literate individuals and consulting practices that understand and leverage these capabilities.

The Authentication Crisis

The power shift dynamics reveal a deeper problem: organizations are losing control of their authentic voice while becoming dependent on artificial enhancement.

When AI Handles Your Company's Voice

Organizations are losing the ability to distinguish human-generated from AI-generated communications within their companies. Decision-makers base strategic choices on artificially enhanced representations of team capabilities rather than authentic assessments.

When AI handles your company's voice, what happens to your company's actual voice? Leadership teams receive strategic documents, analysis reports, and communication materials representing AI capabilities rather than human understanding, creating false impressions of organizational competency.

This narrative becomes dangerous when strategic decisions rely on assumed human capabilities representing AI assistance. Project assignments, role evaluations, and strategic planning suffer from a fundamental misunderstanding of actual vs. enhanced team capabilities.

The Detection Dilemma

The irony deepens when organizations deploy AI detection tools like JustDone.com to identify AI-generated content within their communications. We've created a technological arms race in which AI writes content, humans deploy AI to detect AI content, and then different AI is used to make AI content appear more human.

This layering effect reveals something profound about human psychology: We want the efficiency of AI but crave the authenticity of human connection. The question isn't whether we can detect AI—it's whether detection matters if the communication serves its purpose effectively.

The Human Psyche Factor

What's happening psychologically is even more complex. When people suspect they're reading AI-generated content, they subconsciously discount its value, even if the information is accurate and useful. We're hardwired to seek human connection and an authentic voice. This dilemma creates a paradox: AI can improve communication quality, but human perception of AI involvement can diminish message impact.

The most successful approach combines AI enhancement with authentic human perspective—using AI as a sophisticated editing and research tool while preserving genuine human insight and voice.

The Dependency Trap

Organizations develop critical dependencies on AI for essential business communication without establishing governance frameworks or quality controls. Like any automation technology, overdependence without maintaining underlying skills creates vulnerabilities.

When AI systems fail or provide incorrect information, communication chains become disrupted if staff haven't preserved independent capabilities. Reflecting historical technology dependencies, humans often turn to pen and paper when email servers fail. However, this disruption highlights the underlying risks of dependency.

We're approaching a more profound shift with voice-based AI becoming mainstream. Young professionals already bypass the struggle of learning to write well, jumping directly to AI-assisted communication. This trend creates a generation that can direct AI to communicate for them but cannot communicate effectively themselves.

The Customer Relationship Revolution

The communication transformation extends beyond internal correspondence—revolutionizing how businesses build and maintain customer relationships.

Beyond Communication: AI as Relationship Architecture

In my POS business, communication transformation has evolved into more sophisticated—AI-powered customer relationship orchestration that transforms how restaurants build community and loyalty.

Consider this progression: A restaurant historically sent generic promotional emails. Now, AI analyzes dining patterns, seasonal preferences, and social behavior to create personalized experiences. The system identifies customers who order certain wines and prefer intimate dining, while craft beer customers enjoy group activities.

Seasonal menu changes become targeted community-building opportunities. AI crafts personal invitations that feel individually written: "Sarah, given your preference for our autumn harvest dishes last year, I wanted you to be first to know about our new winter truffle menu launching Tuesday. Would you like your usual corner table for you and Michael?"

The AI agent manages individual customer relationships across multiple touchpoints—direct mail, social media, personal pages, reservation preferences, and dietary accommodations. Each customer receives communication that feels personally crafted because the AI understands their relationship with the restaurant.

Is this customer farming? Absolutely. But it's customer farming that creates genuine value—better experiences, stronger community connections, and more satisfying dining relationships. The technology enables personal attention, which is only possible when restaurant owners know every customer personally.

This new platform represents the platform economy in a microcosm: AI amplifies human relationship-building capabilities, creating experiences that exceed traditional human-only service and pure automation alternatives.

What we're witnessing extends far beyond communication enhancement. AI is becoming the relationship architecture that enables businesses to operate at a human scale while serving thousands of customers.
The restaurant example reveals the broader transformation: AI doesn't replace human connection—it makes authentic human connection scalable. The maitre d' who remembered every customer's preferences now becomes an AI system that enables the entire staff to provide that level of personalized service.

This is the foundation of the platform economy: technology that amplifies human capabilities rather than replacing human value. The most successful businesses will combine AI's analytical and coordination capabilities with human creativity, empathy, and authentic relationship building.

The Generational Communication Shift

While organizations struggle with authentication and dependency, a more fundamental transformation occurs among younger professionals.

The most concerning long-term trend involves young professionals bypassing writing skill development through immediate AI assistance. Voice-based AI enables complete circumvention of written communication learning, creating what I predict is "the beginning of the end of written communication skills for entire generations."

Strategic implications include:

Skill Atrophy:
Professionals who never develop underlying communication capabilities.

Crisis Vulnerability:
When AI assistance disappears, communication effectiveness collapses.

Competitive Disadvantage:
Organizations dependent on AI-assisted communication struggle when technology fails.

Authenticity Loss:
Human perspective and judgment become separated from communication output.

Strategic Response Framework

Understanding these communication transformations requires strategic action at both organizational and individual levels. Success depends on managing AI integration while preserving authentic capabilities.

For Business Leaders

1. Conduct Communication Audits

- Review organizational correspondence from the past 18 months.
- Identify sudden shifts in writing quality, complexity, or style patterns.
- Map changes against known AI adoption within teams.

- Distinguish AI enhancement from AI dependency across different roles.

2. Establish Governance Frameworks

- Develop policies for AI-assisted communication that preserve human skill development.
- Create quality standards that ensure an authentic organizational voice.
- Implement verification processes for critical communications.
- Balance AI efficiency gains with communication skill preservation.

3. Develop Strategic AI Integration

- Train teams on effective AI enhancement rather than AI dependence.
- Establish communication standards that combine AI capabilities with human judgment.
- Create systems for maintaining individual voice while leveraging AI capabilities.
- Invest in AI literacy that preserves rather than replaces human communication skills.

For Individual Professionals

1. Strategic AI Enhancement

- Use AI to improve communication effectiveness rather than replace communication thinking.

Practical Prompt Engineering for Authentic Communication

The key to maintaining an authentic voice while leveraging AI enhancement lies in sophisticated, prompt engineering that preserves human characteristics. Most AI-generated text reveals itself through monotonous cadence and low perplexity—predictable sentence lengths and uniform structural patterns that lack human variability.

Consider this enhanced prompt framework:
"Compose a response in which each sentence contains between 12 and 18 words. Write in an unmistakably human style—natural, conversational, and varied. Use contractions, avoid jargon, and include occasional idiomatic expressions. Here is an example of the

style I seek: [insert example]. Follow this approach, and ensure that every sentence adheres to the specified word count while maintaining the warmth and unpredictability of genuine human prose."

Such prompts invite AI to produce text that combines structural authenticity with human communication patterns...
Maintain the ability to communicate effectively without AI assistance

- Develop frameworks for combining AI capabilities with personal voice and perspective.
- Practice communication skills independently to prevent dependency-related skill atrophy.

2. Competitive Positioning

- Leverage AI enhancement for improved productivity and quality.
- Understand AI communication patterns for competitive intelligence.
- Develop expertise in AI-enhanced communication for consulting and advisory opportunities.
- Position AI literacy as professional development rather than technology dependence.

3. Platform Economy Preparation

- Develop communication quality that meets platform economy participation standards.
- Maintain an authentic voice that distinguishes your contributions from AI-generated content.
- Build communication skills that enable thought leadership and community building.
- Create communication frameworks that demonstrate strategic thinking enhanced by AI.

Quality Control and Authenticity Standards

Sustainable AI integration requires frameworks for maintaining human communication capabilities alongside AI assistance. The goal is AI enhancement that amplifies human thinking rather than replaces human communication development.

Distinguishing Enhancement from Dependence

Enhancement Indicators:

- AI improves the clarity, structure, and professionalism of existing human ideas.
- Individuals can communicate effectively without AI assistance when necessary.
- AI assists strategic thinking rather than replaces strategic thinking
- Communication maintains individual voice and perspective enhanced by AI capabilities.

Dependence Indicators:

- AI generates ideas, structure, and content without significant human input.
- Individuals struggle to communicate effectively without AI assistance.
- AI replaces strategic thinking rather than enhances strategic thinking.
- Communication becomes indistinguishable from AI-generated content.

Platform Communication Standards

Platform economy participation requires communication quality demonstrating human insight enhanced by AI capabilities rather than AI content delivered through human channels.

Platform Contribution Standards:

Authenticity:

Communications reflect genuine human experience and perspective.

Strategic Thinking:

Analysis demonstrates human judgment enhanced by AI research.

Individual Voice:

Content maintains personal communication style and approach.

Value Creation:

Insights provide unique value beyond AI-generated alternatives.

The most successful platform participants combine sophisticated AI enhancement with distinctive human perspectives, creating communications that exceed traditional human capabilities and pure AI alternatives.

The Competitive Implications

The communication revolution creates new forms of competitive advantage while exposing organizations to unprecedented vulnerabilities.

Market Differentiation Through Communication Excellence

Organizations that master strategic AI communication integration gain competitive advantages through superior client interaction, faster response capabilities, and enhanced analytical sophistication. However, success requires governance that preserves authenticity while leveraging AI capabilities.

Companies that struggle will discover they've outsourced their voice to algorithms without realizing it, creating dependency vulnerabilities and competitive disadvantages when authentic communication becomes essential.

Professional Development Imperatives

Individual professionals must develop AI enhancement capabilities that amplify rather than replace communication competency. The most successful approach treats AI as sophisticated support for human strategic thinking rather than a replacement for communication skill development.

Career advancement increasingly depends on combining AI literacy with maintained human communication excellence. Platform economy opportunities favor professionals demonstrating strategic thinking enhanced by AI rather than AI dependence disguised as professional capability.

Communication transformation is just one visible sign of the possible business transformation. Let me show you how to build your entire AI empire, step by step…

The Authentication Strategy Framework

Enhanced Prompt for Authentic Communication:

"Write a response that sounds naturally human with varied sentence length and conversational tone. Use the following guidelines:

- **Sentence Length:** Vary between 8-20 words per sentence
- **Style:** Conversational, using contractions and natural expressions
- **Structure:** Avoid repetitive patterns, include natural transitions
- **Voice:** [Insert example of your writing style]
- **Purpose:** [Specific goal of communication]
- **Audience:** [Specific recipient and their context]

Ensure the response feels like human conversation enhanced by clear thinking, not AI-generated content.

Communication Quality Control Checklist:

Enhancement Indicators (✓ Good):

- AI improves clarity and structure of your existing ideas
- You can communicate effectively without AI when necessary
- Your individual voice and perspective remain evident
- Strategic thinking comes from you, enhanced by AI research

Dependence Indicators (⚠ Warning):

- AI generates ideas and content without significant human input
- You struggle to communicate effectively without AI assistance
- Your communication becomes indistinguishable from AI-generated content
- AI replaces rather than enhances your strategic thinking

Platform Communication Standards:

For Thought Economy participation, your communications must demonstrate:

- **Authenticity:** Reflects genuine human experience and perspective

- **Strategic Thinking:** Shows human judgment enhanced by AI research
- **Individual Voice:** Maintains personal communication style
- **Value Creation:** Provides unique insights beyond AI-generated alternatives

Key Takeaways:

- **Three Transformation Patterns:** Sophistication Spike, Homogenization Effect, Complexity Explosion reveal AI adoption
- **Power Shift Dynamics:** AI-equipped staff can outperform experienced managers through enhanced analytical capabilities
- **Authentication Crisis:** Organizations lose ability to distinguish human from AI-generated communications
- **Strategic Response:** Master AI enhancement while preserving authentic human voice and strategic thinking capabilities

Chapter 15: Building Your AI Empire: From Concept to Platform

Practical Roadmap for AI Business Transformation

But first, let me share something profound about how empire-building begins. The real power isn't in the tools themselves—it's in understanding the fundamental principle that makes AI possible across every domain.

The Universal AI Principle: From Words to Worlds

Something fascinating happens behind the scenes when you type a message to an LLM. The AI doesn't actually "read" your words like humans do. Instead, it converts every word into numbers—mathematical representations called tokens. "Hello" becomes a specific number, "world" becomes another number, and so on.

It may seem minor, but it's crucial for unlocking AI's potential across various fields.

Why This Changes Everything

Here's the breakthrough insight: **If you can convert something into tokens (mathematical representations), you can have AI operate and process it.**

Think about what this means:

- What are your restaurant's sales patterns? Convert them to tokens, and AI can predict busy periods.
- Your customer service conversations? Tokenize them, and AI can identify satisfaction patterns.
- Your manufacturing sensor data? Turn it into tokens, and AI can predict equipment failures.
- Your financial transactions? Tokenize them, and AI can detect fraud patterns.

The Token Revolution: Why AI Can Transform Any Field

Most people think AI is just about text because they see LLMs converting words into tokens. However, this same tokenization concept applies to any AI or ML model. Whether it's:

- Images (pixels as tokens).
- Music (notes and frequencies as tokens).
- Financial data (transactions and patterns as tokens).
- Manufacturing processes (sensor data as tokens).
- Customer behavior (actions and preferences as tokens).
- Medical diagnostics (symptoms and test results as tokens).

When you understand how to tokenize your domain, you unlock AI's ability to process, analyze, and enhance that field.

Empire-building through AI isn't just about writing or coding. AI partnerships can tokenize and enhance every industry, process, and knowledge work.

The Thought Exchange Revolution: The Foundation of All AI Empires

You can take any subject or thought and start a conversation with an LLM to amplify your ideas, expand your vision, and enhance your knowledge. It serves as a foundation for exchanging ideas. Humans possess valuable thoughts based on genuine needs. AI doesn't have this ability yet but excels at building and amplifying ideas. That's how I developed the thought exchange concept.

I have ideas all the time—my brain doesn't switch off. I find myself creating documents to record my thoughts, but later, when I want to join up thoughts or build them out, I spend 25% of my time just locating my previous insights. So, I started developing a "thought engine" using several technologies. It was like an operating system for thoughts. I got the basic parts working to prove the concept, but being time-poor, the project sits waiting for the next major push to make it scalable.

This experience taught me something crucial: **building an AI empire isn't about replacing your thinking—it's about amplifying your unique human insights and making them discoverable, scalable, and valuable.**

This statement highlights how AI can enhance and empower our human capabilities.

The Bidirectional Learning Revolution: Your Thinking Partner Evolution

Here's what most people miss about LLMs: they're not just tools—they're **thinking partners that learn your patterns while teaching you to think more strategically**. This bidirectional symbiosis is the secret foundation of every AI empire, and it's a journey you're an active participant in.

The Learning Loop That Changes Everything:

1. **You teach the LLM your context** through prompts and conversation.
2. **The LLM amplifies your thinking** with enhanced frameworks and connections.
3. **You refine your prompting** based on what produces breakthrough insights.
4. **The LLM adapts to your style** and becomes increasingly effective.
5. **You both evolve** into a more powerful thinking system.

Just tell your LLM what you want, what you see working, and what you don't want. I went through a stage where I was mind-blown by the LLM approach, but it can also work the other way—**the LLM can be mind-blown when you start thinking outside the square**. It's a symbiotic relationship in this context: humans have pure thought, and AI supports your dreams.

Why Good Prompts Create Bidirectional Learning:

The secret most people don't understand is that **good prompts teach you as much as they teach the AI**. When you learn to prompt effectively, you're learning to:

- Articulate your vision with precision.
- Break complex problems into systematic components.
- Consider utilizing frameworks that can be expanded or built upon.
- Recognize patterns in your thinking process.

The LLM becomes your thinking mirror, amplifying your ideas while helping you discover the structure of your thought processes. Here's how to become a master orchestrator rather than merely a tool user.

The Symbiotic Relationship Model: Why AI Needs Your Human Insights

Humans provide:

- Real-world context and lived experience.
- Emotional intelligence and intuitive leaps.
- Ethical judgment and values-based decisions.
- Creative connections across unrelated domains.
- Problem identification based on actual needs.

AI provides:

- Systematic analysis and pattern recognition.
- Rapid processing of complex information.
- Consistent application of frameworks.
- Memory and retrieval of past insights.
- Scaling of thinking across multiple scenarios.

When these capabilities combine through bidirectional learning, you create something neither humans nor AI could achieve: **accelerated wisdom development**.

This chapter provides the step-by-step framework for building your entire AI empire—from personal skill development through platform participation to organizational transformation leadership. But remember: **communication excellence is just the foundation; the real power comes from mastering this bidirectional thinking partnership that makes all other AI capabilities possible.**

Assessment Framework: Where to Start Your AI Journey

The Vessel Tracking Vision: Why Assessment Matters

Let me share one of my favorite projects because it demonstrates the importance of systematic assessment before diving into AI implementation. Three years ago, I designed a concept for a shipping company to monitor five vessels in real-time working in the waters of the Philippines. The idea was elegant: connect GPS to a small PC, create a connecting app, and forward communications from boats to shore via Starlink. At the head office, a small server would collate vessel locations and display them on a map, allowing shared manifest documents and ship logs.

However, I ended up waiting almost two years for Starlink's availability. The project is still pending, but I recently built out further functionality using APIs and geofencing for vessels at sea and anchor. Even though I was early on the technology, there's significant ROI potential through tracking tides and winds with machine learning to reduce fuel costs.

This project taught me that being visionary isn't enough. You need a systematic assessment to understand technology readiness, infrastructure requirements, and implementation timing. Like any project using AI to specify requirements, **it requires you to learn to visualize technology and apply AI to enhance your thinking or the customer's vision to achieve accuracy and monitoring capabilities.**

It's about working through the possibilities and brainstorming what should be in development, in Phase 2, or what remains a dream. I live for this part of consulting: pushing boundaries and discovering innovative perspectives, with AI helping to shape those ideas.

The Four-Dimension Assessment Model

Before implementing any AI transformation, you need a systematic assessment that prevents overwhelm and misdirection. My consulting methodology uses four critical dimensions to determine your optimal starting point and implementation strategy.

Dimension 1: Personal/Team AI Literacy

Current skill levels across the four mastery levels we explored in Chapter 4 determine your learning pathway and timeline requirements. Assessment includes:

- **Level Assessment:** Which of the four AI mastery levels describes your current capability?
- **Learning Capacity:** Available time for skill development and experimentation.
- **Resistance Patterns:** Individual or team concerns about AI adoption require change management.
- **Support Systems:** Access to training resources, mentorship, and implementation guidance.
- **Bidirectional Learning Readiness:** Ability to engage in collaborative thinking partnerships with AI.

Dimension 2: Business Process Analysis

Identifying high-value AI implementation opportunities requires understanding which processes benefit most from AI enhancement:

- **High-Volume Activities:** Rule-based tasks suitable for immediate AI automation.
- **Customer-Facing Processes:** Personalisation and service enhancement opportunities.
- **Decision-Making Workflows:** Analysis requirements that benefit from AI-enhanced insights.
- **Knowledge Management:** Information processing and distribution improvements.
- **Creative Problem-Solving:** Areas where human-AI collaboration creates breakthrough innovations.

Dimension 3: Market Positioning Evaluation

Strategic positioning requires understanding competitive dynamics and customer expectations:

- **Competitive Landscape:** AI adoption levels among direct competitors and industry leaders.
- **Customer Readiness:** Market expectations and acceptance of AI-enhanced services.
- **Differentiation Opportunities:** Unique AI capabilities that create competitive advantages.

- **Industry Evolution:** Regulatory, technical, and market trends affecting AI implementation.
- **Thought Leadership Potential:** Opportunities to become recognized for AI innovation in your domain.

Dimension 4: Resource and Infrastructure Assessment

Implementation success depends on a realistic evaluation of available resources and technical capabilities:

- **Budget Allocation:** Investment is available for AI tools, training, and implementation.
- **Technical Infrastructure:** Current systems' ability to integrate AI capabilities.
- **Organizational Support:** Leadership commitment and cultural readiness for change.
- **Timeline Constraints:** Business cycles and project schedules affecting implementation timing.
- **Platform Preparation:** Infrastructure needed for thought economy participation.

The Assessment Methodology

My proven assessment approach prevents the overwhelm that destroys most AI initiatives. Instead of trying to transform everything simultaneously, systematic assessment identifies your highest-value starting points.

Step 1: Current State Mapping

Document existing processes, skill levels, and resource availability across all four dimensions. Establishing a foundational understanding highlights the priorities for implementation.

Step 2: Opportunity Identification

Analyse the assessment results to identify quick wins, strategic advantages, and transformation possibilities that align with your specific context.

Step 3: Readiness Evaluation

Determine technical, cultural, and strategic preparation requirements before implementation.

Step 4: Pathway Selection

Choose the implementation sequence and timeline based on assessment results rather than generic best practices.

Step 5: Bidirectional Learning Setup

Establish systems for capturing and building upon the insights generated through human-AI collaboration.

The Boston Dynamics Warning: Why Speed Matters

Here's why assessment can't become analysis paralysis. A business's size is often its downfall regarding AI adoption. Looking at established companies, you can see their dreams at play and easily layer their complexity and capital invested over the years to build dominance in a product, market, or both.

Introducing new ideas is treated with caution, and rightly so—they've invested heavily in intellectual property to carve out market share. However, not understanding the profound impact that AI is accelerating and not learning to quantify AI so their business can solve problems or rapidly adopt it is almost a crime for founders who once saw market gaps and started businesses.

Let me illustrate with Boston Dynamics versus the new generation of robotics companies.

Boston Dynamics spent years developing Atlas through traditional engineering methods—hand-coding locomotion and movement patterns. Meanwhile, newer robotics companies use AI-driven approaches with digital twin simulations and parallel training environments. While these new methods haven't yet achieved the full complexity of Atlas in 100 days, they're showing remarkable progress in specific tasks. Recent research shows robots like Unitree H1 and others can learn complex behaviors like box carrying in just 2-6 hours of simulation time.

The acceleration is real, even if not as dramatic as initial claims suggested. Here lies my point: **existing, well-placed large companies are not safe.** The typical AI-first approach follows: CAD design → digital twin simulation (using tools like NVIDIA Isaac Sim) → rapid iteration through thousands of virtual training cycles → deployment.

Such an approach enables startups to operate quietly in the market. When traditional businesses recognize this shift, they will struggle to catch up. One of my key messages is that companies must prioritize AI and evolve into technology leaders, producing X and Y. We're on the verge of seeing a single person become the world's first trillionaire through a startup—this will only be achievable through AI.

The business traditionally looks for ROI on technology, like in past years when we adopted technology at a steady rate year on year. The formula is proven—why change? "Show me the ROI on AI," they say. Then tell me when the competitors are using AI."

You'll react too late once you discover what your competitors are doing—they will eat your lunch before you know it.

The Enterprise AI Reality

I must acknowledge that the heavy hitters are present for larger ventures—it's not all about bottom-end AI tools. You can step into companies like C3 and Palantir Technologies and then into AWS, Google, and Microsoft Solutions. However, AI-ready companies like C3 and Palantir can apply their business models to integrate AI with a human-in-the-loop approach.

Companies like DataRobot, H2O.ai, Scale AI, Dataiku, Cognite, and Veritone have architectural advantages because they don't need to retrofit human oversight into existing systems—it's baked into their DNA. They understand that **AI augments rather than replaces human expertise**, which is crucial for enterprise adoption and regulatory compliance.

We've been focusing on AI LLM integration, but these companies are the corporate AI leaders with robust solutions. Palantir, for instance, is used by the US military. The key differentiator is that their platforms are designed for human-AI collaboration from day one, not as an afterthought.

A three-tiered reality emerges: personal AI expertise, corporate AI platforms, and entrepreneurial AI disruption. Your strategy for building an empire should actively incorporate all three elements.

The Phased Implementation Timeline

Phase 1: Foundation Building (Months 1-3)

The first 90 days establish fundamental AI literacy and identify immediate value opportunities without disrupting existing operations.

Month 1: AI Literacy Development

- **Personal Skill Assessment:** Determine current AI mastery level using Chapter 4 Framework
- **Tool Familiarisation:** Begin using basic AI tools for everyday tasks in your domain
- **Learning System Setup:** Establish regular learning schedule and progress tracking
- **Quick Win Identification:** Find 2-3 processes that benefit immediately from AI assistance
- **Bidirectional Learning Initiation:** Begin developing your personal prompting style and AI collaboration patterns

Month 2: Process Integration

- **Workflow Enhancement:** Integrate AI tools into daily work routines
- **Quality Development:** Learn to evaluate and improve AI-generated output
- **Collaboration Patterns:** Develop effective human-AI working relationships
- **Measurement System:** Track productivity improvements and quality changes
- **Thinking Partnership Development:** Begin recognizing how AI amplifies your unique insights

Month 3: Strategic Expansion

- **Advanced Applications:** Move beyond basic tool usage to strategic AI thinking
- **Team Integration:** Share AI capabilities with colleagues and gather feedback
- **Client Value:** Begin delivering AI-enhanced value to customers or stakeholders

- **Platform Exploration:** Investigate AI communities and collaborative opportunities
- **Thought Capture Systems:** Implement methods for capturing and building upon AI-enhanced insights

Foundation Phase Success Metrics:

- 2-5x productivity improvement in specific tasks
- Measurable quality enhancement in work output
- Comfortable daily AI usage without dependence concerns
- A clear understanding of AI capabilities and limitations
- **Evidence of bidirectional learning:** Improved prompting skills and AI responsiveness to your style

Phase 2: Strategic Implementation (Months 4-12)

The strategic phase focuses on comprehensive business transformation and competitive advantage development.

Months 4-6: Business Transformation

- **Process Redesign:** Reimagine core workflows with AI enhancement as foundational
- **Service Innovation:** Develop new capabilities and offerings enabled by AI
- **Competitive Positioning:** Leverage AI capabilities for market differentiation
- **Team Development:** Expand AI literacy across your organization or network
- **Thought Engine Development:** Build systems for scaling your AI-enhanced insights

Months 7-9: Market Leadership

- **Industry Expertise:** Develop specialized AI applications for your domain
- **Thought Leadership:** Share insights and frameworks through content and speaking
- **Network Building:** Connect with other AI-enhanced professionals and organizations

- **Platform Participation:** Begin contributing to AI communities and collaborative projects
- **Symbiotic Relationship Mastery:** Achieve seamless human-AI collaboration in complex projects

Months 10-12: Platform Preparation

- **Community Building:** Develop a following around your AI expertise and insights
- **Revenue Diversification:** Create income streams beyond traditional services
- **Platform Strategy:** Prepare for thought economy participation and leadership
- **Strategic Partnerships:** Build relationships that enable platform economy engagement
- **Thought Trading Preparation:** Develop valuable frameworks and insights for community exchange

Strategic Phase Success Metrics:

- Fundamental business model enhancement through AI capabilities
- Recognition as an AI leader within your industry or domain
- Multiple revenue streams enabled by AI expertise
- Active participation in AI communities and professional networks
- **Demonstrated ability to generate unique insights through human-AI collaboration**

Phase 3: Platform Leadership (Months 12+)

The leadership phase transitions from AI user to platform participant and thought economy contributor.

Platform Development Strategy:

- **Technical Infrastructure:** Develop systems for thought trading and community engagement
- **Content Creation:** Produce valuable insights and frameworks for platform communities
- **Community Recruitment:** Build an audience for your expertise and collaborative projects

- **Revenue Model Implementation:** Monetise platform participation through multiple streams
- **Bidirectional Learning at Scale:** Help others develop effective AI collaboration patterns

Future Vision Integration:

- **Continuous Innovation:** Stay current with AI advancement and maintain a competitive edge
- **Strategic Positioning:** Establish a long-term market position in the evolving AI economy
- **Community Leadership:** Take governance and development roles in platform communities
- **Passive Income Development:** Build sustainable revenue streams through platform participation
- **Thought Economy Contribution:** Become a recognized contributor to the global exchange of AI-enhanced insights

The ROI Revolution: POS Business Transformation

I give you a concrete example of ROI thinking from my POS business experience. The typical business structure is where restaurants purchase stock when they get low. AI can change this completely.

Labor costs about 30% of operational costs in a restaurant. A rethink with automation could save 10%, and with AI being so cheap to deploy and run, the ROI will be months, not years. To become AI-centric, you must adopt a bold and uncompromising approach.

Don't think of older ways of doing business—becoming AI-first and using AI to deliver the new way is simpler. I hear people scream about this approach. It's like the best softball pitcher pitching to the best batter: one knows the next move and has to react.

The key insight: When you develop a bidirectional learning relationship with AI, you don't just automate existing processes—you reimagine entire business models. The AI helps you see possibilities you couldn't envision alone, while your industry knowledge lets the AI understand which innovations will work in practice.

Investment Requirements and Benefit Measurement

Short-Term Investment (Months 1-6):

- **Training and Education:** $2,000 → 5,000 for comprehensive AI literacy development
- **Tool Subscriptions:** $100 → 500 monthly for professional AI platforms and services
- **Implementation Time:** 10 → 20 hours weekly for learning and integration
- **Consulting Support:** $5,000 → 15,000 for accelerated transformation guidance
- **Bidirectional Learning Development:** Additional 5 → 10 hours weekly for developing AI collaboration skills

Note: These are personal/organizational development costs. Once mastered, AI-enhanced professionals can deliver services at 80-95% cost reduction compared to traditional methods (see Chapter 3 examples).

Medium-Term Investment (Months 6-18):

- Advanced Tools: $500 → 2,000 monthly for sophisticated AI platforms and integrations
- Infrastructure Development: $10,000 → 50,000 for custom AI applications and systems
- Team Training: $5,000 → 25,000 for organizational AI literacy development
- Market Positioning: $10,000 → 30,000 for thought leadership and platform participation
- **Thought Engine Development:** $5,000 → 20,000 for systems that capture and scale AI-enhanced insights

Expected Returns Timeline:

- **Month 1-3:** 200% → 500% productivity improvement in specific tasks
- **Month 4-6:** 20 → 50% overall business efficiency improvement
- **Month 7-12:** 100 → 300% revenue increase through new capabilities and market positioning

- **Year 2+:** 200 → 500% total business value increase through platform economy participation
- **Ongoing:** Compound returns from bidirectional learning relationships that continue improving over time

My Platform Evolution Journey

Let me share how I evolved my platform strategy because the progression teaches essential lessons about building versus scaling.

I started small—copying and pasting using AI—then progressed to Windsurfer, Firebase, and Claude Code. Google Firebase was a bit too early for me and slowed me down because it was bleeding-edge, but the scalability was there.

Here's the key insight: **If you have an idea, consider whether you're building a concept or building to scale with AI. Try creating the latter, as AI builds core structure so quickly.**

When I moved to Claude Code, I discovered something profound about the relationship between human vision and AI execution. **The LLM can be mind-blown when you start thinking outside the square.** This symbiotic relationship—humans providing pure thought, AI delivering support for your dreams—becomes the foundation of platform economy participation.

The Evolution of Bidirectional Learning:

1. **Tool Usage Phase:** Using AI to execute your ideas faster
2. **Collaboration Phase:** AI begins suggesting improvements to your ideas
3. **Symbiosis Phase:** You and AI create ideas neither could generate alone
4. **Amplification Phase:** Your combined thinking capability scales beyond individual limits
5. **Platform Phase:** You teach others to develop their bidirectional learning relationships

Each phase builds upon the previous one, but the magic happens when you realize **you're not just using AI—you're evolving a thinking partnership that becomes more valuable over time.**

Platform Economy Transition: From User to Leader

The ultimate goal of building an AI empire is transitioning from tool user to platform economy leader. Your value arises not from using AI but from generating unique insights through mutual learning.

Early Platform Participation:

- Active contribution to AI communities and professional networks
- Regular publication of AI insights and strategic frameworks
- Collaboration requests from other AI-enhanced professionals
- Speaking opportunities at industry events and professional conferences
- **Demonstrated mastery of bidirectional learning:** Teaching others how to develop effective AI collaboration

Platform Leadership Development:

- Community governance roles and collaborative project leadership
- Multiple revenue streams from platform economy participation
- Recognition as a thought leader in AI business transformation
- Sustainable passive income from platform-based intellectual property
- **Thought Economy Contribution:** Creating frameworks that help others develop their bidirectional learning capabilities

The Future Integration: Continuous Evolution

Technology Evolution Tracking:

- Capability Monitoring: Stay current with AI advancement and new platform features
- Integration Opportunities: Identify new applications for emerging AI capabilities
- Competitive Intelligence: Understand how AI development affects your industry and market
- **Bidirectional Learning Evolution:** Adapt your collaboration patterns as AI capabilities advance

Strategic Positioning Maintenance:

- Thought Leadership: Continue developing and sharing insights about AI transformation
- Community Engagement: Maintain active participation in platform economy communities
- Capability Development: Expand AI expertise into new domains and applications
- **Relationship Deepening:** Continuously improve the quality and value of your human-AI collaboration

Long-Term Platform Economy Participation:

- Governance Roles: Take responsibility for platform community development and standards
- Mentorship Programs: Help other professionals navigate AI transformation successfully
- Framework Development: Create methodologies that improve community-wide AI implementation
- **Bidirectional Learning Teaching:** Help others master the art of collaborative thinking with AI

The Final Integration: Your Empire Awaits

You now have the complete implementation roadmap, from the initial assessment to platform leadership development. But remember: **everything starts with mastering the bidirectional learning relationship between you and AI.**

It's not only about using tools more effectively; it's about building a thinking partnership that enhances your unique human insights while teaching you to think more strategically and systematically. **When you master this relationship, you don't just build an AI empire—you become a leader in the thought economy.**

The next chapter reveals the ultimate insight that makes this entire book not just an analysis of AI transformation but an invitation to join the creation of the thought economy. You've been participating in this economy from the moment you started reading.

The 90-Day Quick Start Program

Days 1-30: Foundation Building

- Week 1: AI Literacy Assessment + Tool Setup + First Applications
- Week 2: Workflow Integration + Quality Development + Productivity Measurement
- Week 3: Strategic Expansion + Team Introduction + Client Value Creation
- Week 4: Platform Exploration + Community Engagement + Progress Review

Days 31-60: Strategic Implementation

- Week 5-6: Process Redesign + Service Innovation + Competitive Positioning
- Week 7-8: Team Development + Industry Expertise + Network Building

Days 61-90: Platform Preparation

- Week 9-10: Community Building + Revenue Diversification + Platform Strategy
- Week 11-12: Strategic Partnerships + Thought Leadership + Future Planning

Investment vs. Return Timeline

Phase	Investment	Expected Returns	Key Milestones
Month 1-3	$2,000-$5,000	50-200% productivity gains	Daily AI usage, measurable improvements
Month 4-6	$10,000-$25,000	100-300% revenue increase through new AI-enhanced capabilities	New capabilities, market recognition
Month 7-12	$25,000-$75,000	200-500% total business value increase through platform economy participation	Platform participation, thought leadership
Year 2+	Platform participation	Compound returns	Community leadership, passive income

The Bidirectional Learning Framework

Phase 1: Tool Usage - Using AI to execute your ideas faster
Phase 2: Collaboration - AI begins suggesting improvements to your ideas
Phase 3: Symbiosis - You and AI create ideas neither could generate alone
Phase 4: Amplification - Combined thinking capability scales beyond individual limits **Phase 5: Platform Leadership** - Teaching others to develop AI collaboration relationships

Key Takeaways:

- **Universal AI Principle:** Everything that can be tokenized (converted to mathematical representations) can be enhanced by AI
- **Bidirectional Learning:** AI doesn't just follow your directions—it learns your patterns while teaching you strategic thinking

- **Implementation Timeline:** 90-day quick start → 12-month strategic development → Platform leadership
- **ROI Revolution:** Bold AI-first approaches create months (not years) payback periods with exponential scaling potential

The plot is about to be revealed, and your role in this transformation is about to become clear.

Chapter 16: The Plot Revealed: You've Been Trading Thoughts All Along

The Meta-Revelation That Transforms Readers Into Participants

Your Participation Invitation

The Recursive Loop Completes:

- 1983: Gödel, Escher, Bach taught me about self-reference and strange loops
- 1984: I tried to build AI systems that could think about their thinking
- 2025: I wrote a book about the Thought Economy using thought economy methods
- Today: You're living the theory by trading your consciousness for strategic intelligence

What Just Happened:

Every framework you absorbed, every insight you gained, every strategic connection you made—that was commerce. You traded your attention and cognitive engagement for knowledge with immediate market value.

The Formation Begins Now:

We're seeking three types of partners to build the infrastructure before tech giants realize what we're creating:

Type 1: Thought Producers

Your daily observations have economic value

What You Contribute:

Raw insights from lived experience

- "This process is inefficient" = $50 insight
- "Customers struggle with this feature" = $200 framework contribution
- "This trend connects to that opportunity" = $1,000 strategic analysis

What You Receive:

Revenue sharing from insights that create business value

Your Next Step:

Begin documenting daily observations about inefficiencies, opportunities, and insights in your domain

Type 2: Thought Farmers

Strategic thinkers who cultivate insights with AI partnership

What You Contribute:

Transform raw observations into valuable frameworks

- Organize individual perspectives into scalable methodologies
- Use AI partnership to develop insights neither humans nor AI could create alone
- Create platforms where thought producers and buyers can connect efficiently

What You Receive:

Revenue from orchestrating intelligence + platform ownership equity

Your Next Step:

Master the bidirectional learning framework from Chapter 15 and begin building AI collaboration capabilities

Type 3: Thought Capital

Resources to scale before corporate incumbents enter the market

What You Contribute:

Investment in platform development and global expansion

- Platform technology that matches AI advancement curves

- Marketing to recruit thought producers before competition emerges
- Infrastructure enabling global thought trading at scale

What You Receive:

Equity in platforms that monetize human consciousness

Your Next Step:

Connect with us to discuss investment opportunities in thought economy infrastructure

The Three Critical Choices:

Option 1:

Wait and become consumers in AI-dominated intelligence markets

Option 2:

Join the thought farming revolution while humans can still direct the partnership

Option 3:

Lead by providing thought capital to ensure humans share economic returns from consciousness monetization

Your Immediate Action Plan:

If You're a Thought Producer:

1. Begin Chapter 4's AI Mastery assessment today
2. Start documenting insights from your daily work
3. Join our community platform to connect with thought farmers
4. Develop your unique perspective into tradable intelligence

If You're a Thought Farmer:

1. Master the frameworks from Chapters 5-6 (Mentor Method + Prompt Engineering)
2. Begin building AI collaboration relationships using Chapter 15's timeline
3. Identify thought producers in your network who need AI amplification

4. Create platforms where others can exchange valuable insights

If You're Thought Capital:

1. Review the investment requirements from Chapter 15
2. Assess the market opportunity in the emerging thought economy
3. Connect with our team to discuss infrastructure development
4. Position yourself to own equity in consciousness monetization platforms

The Calling Is Clear:

The refrigerated hard drive taught me that breakthrough solutions come from impossible connections. The 1984 AI project taught me that human consciousness could be systematized. The 2025 book proves that thoughts become currency when properly cultivated.

Your thoughts are the seeds. AI is the farming tool. The platform is the land. Collaborative intelligence is the harvest.

But the agricultural window is closing as AI accelerates beyond human coordination speed.

The question isn't whether consciousness becomes monetized—it's whether humans will farm the harvest or merely feed the machine.

Key Takeaways:

- The Meta-Revelation: You've been participating in the thought economy from the moment you started reading this book
- The Formation: We're building infrastructure for trading human consciousness as economic currency
- Three Partner Types: Thought Producers (generate insights), Thought Farmers (organize intelligence), Thought Capital (scale platforms)
- The Choice: Join the thought farming revolution while humans can still direct AI partnership, or watch others monetize what you could have owned

CONCLUSION: The Future We're Building Together

The Great AI Enlightenment isn't just about understanding technological change—it's about participating in the creation of a new economy where human consciousness becomes the most valuable currency.

What We've Discovered:

- AI democratizes technical capabilities, making human perspective scarce and valuable
- The Thought Economy emerges as the inevitable result of AI transformation
- Individual AI mastery enables competitive advantages over traditional institutions
- Platform participation creates sustainable revenue from consciousness trading

What Happens Next:

The formation begins now. Every reader who understands these frameworks becomes a founding participant in the thought economy. Your role—whether as Thought Producer, Thought Farmer, or Thought Capital—determines how human consciousness maintains economic value in an AI-accelerated world.

Your Transformation Awaits:

Welcome to the thought farming revolution. Your participation determines whether humans partner with AI or serve it. The future we've been reading about becomes the future we're building together.

Your invitation:

Ride the AI wave as a farmer, not food.

Appendix: Quick Reference Guides

The Four Levels of AI Mastery - Quick Assessment

Level 1 (Curiosity Phase - Months 1-2):

- I use AI occasionally for creative tasks and simple questions
- I'm amazed by AI capabilities but unsure of practical applications
- I sometimes get poor results and don't understand why
- I'm building communication skills that will become valuable in thought-trading communities

Level 2 (Practical Phase - Months 3-6):

- I use AI weekly for work tasks with measurable productivity gains
- **I think of AI as a research team and learning accelerator**
- **Colleagues notice quality improvements in my work**
- **I can synthesize information to create genuine value for others**

Level 3 (Integration Phase - Months 5-10):

- **I've completed technical projects with AI assistance**
- **I can read and modify AI-generated code**
- **I think "What can I build?" instead of "What can I buy?"**
- **I create tools and solutions that others want to use and learn from**

Level 4 (Transformation Phase - Months 8-18):

- **I've designed autonomous systems that operate independently**
- **Others seek my expertise in AI strategy and implementation**
- **I think in terms of "intelligent systems" and "cognitive architecture"**
- **I create value that competitors cannot easily replicate**

The C.L.E.A.R. Prompt Framework - Quick Reference

C - Context Architecture:

Set the stage with specific role, situation, and background

- "Act as a senior market analyst with 15 years of experience in hospitality technology..."

L - Logic Specification:

Define analytical approach and frameworks

- "Use Porter's Five Forces framework combined with PESTEL analysis..."

E - Examples Integration:

Provide reference points and comparisons

- "Reference successful market entries like Square's expansion into European markets..."

A - Action Orientation:

Specify deliverables and next steps

- "Provide specific recommendations with implementation timelines..."

R - Results Refinement:

Optimize output through iteration

- "After providing initial analysis, identify the three weakest assumptions..."

The Three-Player Competitive Game - Quick Reference

Player 1: Traditional Incumbents

- Advantages: Relationships, resources, credibility
- Strategy: Strategic specialization or comprehensive transformation
- Risk: Organizational inertia and legacy constraints

Player 2: AI-Native Companies

- Advantages: Technology-first design, cost structure, scalability
- Strategy: Market segment domination with partnership for complex work
- Risk: Relationship deficits and creative limitations

Player 3: AI-Enhanced Entrepreneurs

- Advantages: Speed, agility, personal relationships
- Strategy: Hybrid model combining AI capabilities with human judgment
- Risk: Resource limitations and scale challenges

Platform Economy Participation - Action Steps

Phase 1: Foundation (Months 1-3)

1. Complete AI literacy assessment using Chapter 4 framework
2. Begin daily AI usage for practical tasks
3. Develop basic prompt engineering skills
4. Identify your unique perspective and domain expertise

Phase 2: Strategic Development (Months 4-12)

1. Master the mentor framework from Chapter 5
2. Build advanced prompt engineering capabilities
3. Create valuable content and frameworks
4. Begin community participation and thought leadership

Phase 3: Platform Leadership (Months 12+)

1. Develop multiple revenue streams from AI expertise
2. Build communities around your insights
3. Create tools and frameworks others want to use
4. Participate in governance of platform communities

Investment Timeline - Expected Returns

Short-term (Months 1-3):

- Investment: $2,000-$5,000 (training, tools, implementation time)
- Returns: 50-200% productivity gains in specific tasks
- Milestones: Daily AI usage, measurable improvements

Medium-term (Months 4-6):

- Investment: $10,000-$25,000 (advanced tools, infrastructure, team training)
- Returns: 100-300% revenue increase through new capabilities
- Milestones: Market recognition, enhanced service delivery

Long-term (Months 7-12):

- Investment: $25,000-$75,000 (platform development, strategic positioning)
- Returns: 200-500% total business value increase
- Milestones: Platform participation, thought leadership

Ongoing (Year 2+):

- Investment: Platform participation and continuous learning
- Returns: Compound returns through platform economy participation
- Milestones: Community leadership, passive income streams

The AI Communication Assessment Tool

Enhancement Indicators (✓ Good):

- AI improves clarity and structure of your existing ideas
- You can communicate effectively without AI when necessary
- Your individual voice and perspective remain evident
- Strategic thinking comes from you, enhanced by AI research

Dependence Indicators (⚠ Warning):

- AI generates ideas and content without significant human input
- You struggle to communicate effectively without AI assistance
- Your communication becomes indistinguishable from AI-generated content
- AI replaces rather than enhances your strategic thinking

Thought Economy Participation Types

Type 1: Thought Producers

- Contribute: Daily observations and insights from lived experience
- Receive: Revenue sharing from insights that create business value
- Next Step: Document observations about inefficiencies and opportunities

Type 2: Thought Farmers

- Contribute: Transform raw observations into valuable frameworks
- Receive: Revenue from orchestrating intelligence + platform equity
- Next Step: Master bidirectional learning with AI partnership

Type 3: Thought Capital

- Contribute: Investment in platform development and scaling
- Receive: Equity in platforms that monetize human consciousness
- Next Step: Connect for investment opportunities in thought economy infrastructure

Quick Start Checklist

Week 1: Foundation

- Complete AI mastery level assessment
- Set up basic AI tools (ChatGPT, Claude, etc.)
- Begin daily AI usage for routine tasks
- Start documenting insights and observations

Week 2: Integration

- Learn C.L.E.A.R. prompt framework
- Apply AI to work projects
- Measure productivity improvements
- Identify unique value creation opportunities

Week 3: Strategic Development

- Master mentor framework basics
- Begin building AI collaboration patterns
- Create first valuable insight or framework
- Connect with AI community

Week 4: Platform Preparation

- Develop thought leadership content
- Build professional network around AI expertise
- Identify platform participation opportunities
- Plan next phase of development

Living Analysis - Real-Time Validation of the AI Revolution

"The predictions in this book continue to unfold in real-time. For the latest analysis and case studies that validate these frameworks, visit:"

www.thought-economy.com/insights

Updated Often with:

- **Case studies of companies implementing AI-native operations**
- **Analysis of new automation announcements**
- **Real-time validation of the economic models presented in this book**
- **Reader discussions and community insights**

Emergency Contact Information

For Technical Support:

- Review Chapter 4 (Four Levels of AI Mastery)
- Consult Chapter 6 (Prompt Engineering Framework)
- Access community forums and resources

For Strategic Guidance:

- Review Chapter 11 (Flower Market of Human Consciousness)
- Consult Chapter 15 (Building Your AI Empire)
- Connect with thought farming community

For Platform Participation:

- Review Chapter 16 (The Plot Revealed)
- Visit community platform
- Connect with other thought farmers and producers

Remember: The goal isn't to replace human capabilities with AI, but to amplify human consciousness through strategic AI partnership. Your unique perspective, enhanced by AI collaboration, becomes valuable currency in the thought economy.

About the Author

Alister Thyne brings nearly five decades of technology implementation experience to the AI revolution. From fixing refrigerated hard drives in 1980 to developing AI-powered POS systems in 2025, he has witnessed and participated in every major technological transformation of the modern era.

As a solutions consultant specializing in enterprise transformations and the founder of multiple hospitality technology companies, Alister has developed unique insights into how businesses adapt to technological change. His experience spans hardware engineering, software development, business consulting, and AI implementation across industries from restaurants to enterprise corporations.

His prediction of robotic restaurants in 2003—dismissed as "too early" by venture capitalists—has become reality in 2025, demonstrating his ability to recognize transformation patterns decades before they materialize. This book represents the culmination of his pattern recognition expertise applied to the most significant technological transformation in human history.

Alister lives in Sydney, Australia, where he continues developing AI-enhanced business solutions while building the infrastructure for the emerging thought economy.

Connect With the Author

Join the Thought Economy Community:

- Website: www.thought-economy.com
- LinkedIn: www.linkedin.com/in/alisterthyne
- Email: info@thought-economy.com

Ready to Start Your Transformation?

- Download the AI Mastery Assessment Tool
- Access the C.L.E.A.R. Prompt Templates
- Join the 90-Day Implementation Program
- Connect with other Thought Farmers and Producers

Investment Opportunities:

- Platform Development Partnership
- Thought Economy Infrastructure Investment
- Early Access to Consciousness Trading Platforms

The formation begins now. Your participation determines the future we build together.

Notes

Made in United States
Orlando, FL
04 November 2026

90201291R00148